KENTISH OASTS

16th-20th CENTURY

THEIR HISTORY, CONSTRUCTION AND EQUIPMENT

KENTISH OASTS

16th-20th CENTURY

THEIR HISTORY, CONSTRUCTION AND EQUIPMENT

by **ROBIN and IVAN WALTON**

with a FOREWORD by RICHARD WOOD

Published By
CHRISTINE SWIFT
BURNT MILL, EGERTON, KENT

Published By
CHRISTINE SWIFT
1998

KENTISH OASTS
16th-20th CENTURY
THEIR HISTORY, CONSTRUCTION
and
EQUIPMENT

By
ROBIN and IVAN WALTON

Printed by
MICKLE PRINT Ltd, WESTMINSTER ROAD, CANTERBURY

Binding by
ROBERT SWIFT, BURNT MILL, EGERTON

1st EDITION CLOTH

ISBN No. 0 9506977 7 X

To our wives Betty and Linda for their patience, understanding and support.

PREFACE

Since the publication of our original definitive book entitled 'Oasts in Kent, 16th - 20th Century' way back in 1985, many changes have taken place in the hop industry throughout Kent.

The number of oasts being used for commercial hop drying has continued to decline, falling from 197 to 72 over these last 12 years and many more of the oasts constructed during the Victorian era have been converted to numerous other uses.

This new publication 'Kentish Oasts' has not only completely updated the former 'Oasts in Kent' book, but has been expanded to include several additional sections and many extra facts, photographs, sketches and diagrams.

The additional material covers the former hop growing area of Kent, now part of the Greater London Area. Also a section traces the emigration of a Kent hop farming family 'down under' to Van Diemans Land (Tasmania), with their unique oast constructions and use of Kent manufactured hop pressing equipment.

Other features include a much enlarged coverage of the Counties Museums and local hopping attractions. Also an ingenious portable hop drying system.

Robin & Ivan Walton,
September 1997.

ACKNOWLEDGEMENTS

The authors express their grateful thanks to

Bromley Central Library, Hobart Library, Kent Archaeological Society, Kent County Library and Centre for Kentish Studies, Museum of Kent Life, Nature Conservation Trust, Orpington Library, Rochester Bridge Trust.

Courage Ltd., Guinness plc, Shepherd Neame Ltd., Truman Ltd., Whitbread plc.

Drake & Fletcher Ltd, English Hops Ltd., Glaxo Wellcome plc, Heath Engineering Works Ltd., Hop Engineering, James Clifford & Son Ltd., Morris Hanbury Ltd.

and the following individuals

C.Amos, T.Ayears, J.Blest, J.Britcher, J.Buggs, R.Chater, N.Clark, R.Coleman, A.Coombe-Jones, A.Cronk, R.Day, C.Dolding, J.Eastland, R.Farrar, T.Frankcomb, J.Gibson, A.Gorst-Williams, V.Guest, S.Highwood, R.Holmes, A.Hulme, C.Lamb, N.Laslett, M.Lawrence, E.Lunato, A.Madle, R.Malins, F.Meakin, P.Morrish, B.Morton, B.Ovenden, R.Pierce, D.Salier, D.Saunders, M.Scott, G.Stilwell, C.Strickland, R.Swift, J.Todd, D.Turner, R.Warner, B.Wenham, L.Wheeler, R.Wood, D.Worley.

We would also like to thank the owners and occupiers of all the many oasts examined in the course of this study.

FOREWORD

The Flemish weavers arrived in Kent early in the 16th century to take advantage of the county's prosperous wool industry. They brought with them new varieties of hops. Brewers in the low countries, having found that the hops had preservative properties that prolonged the life of ale brewed with malt, yeast and water. The need to develop hop drying facilities in Kent, therefore, dates back to the arrival of these early settlers.

The fortunes of Kent hop growers have gone through a number of economic cycles since their introduction. The golden years of hop growing were during the later part of the 19th century, when 72,000 acres (29,000 hectares) were growing in the English countryside and amongst the many varieties introduced at that time was the Fuggle, a new hop named after Mr Richard Fuggle in 1875. Both Fuggles and Goldings, which date back to around the late 1700's, remain today amongst the most sort after varieties by brewers of traditional English Cask Ales in this country, and abroad, where English ales are attracting increasing interest from discerning beer drinkers the world over.

It was in 1774 that an act was passed requiring "Pockets", in which hops were packed, to be stencilled with the grower's name, year and place of growth, a practice that is still a requirement today.

By the early part of the 20th century the golden years of hop growing were over. This was as a direct result of the Government raising the tax on beer from 7 shillings and 9 pence to 23 shillings a barrel. Beer consumption fell, and with it the demand for hops. In September of 1915 hop growers were so concerned that they appealed to the Government to introduce an import duty on foreign hops. In response to continued pressure a ban on all foreign hops by law was instituted in June of 1916. Government controls on beer production in 1916 and 1917 brought additional pressure on hop demand, and it was not until 1932 that market stability returned to the hop trade with the introduction of the Hops Marketing Board (HMB). During the life of the HMB, which was finally dissolved in 1982, a hop quota was in place and this quota balanced supply with the demand from British brewers. A time of prosperity for hop growers returned. Since the dissolution of the HMB, hop growers have once again had to face fierce competition from cheap bittering hops from the USA in particular, but the traditional growers of varieties such as Fuggles and Goldings have been able to enjoy an increasing demand for their unique hop aroma. Since their introduction into this country in the 16th century tradition has been a very important factor in the production and trade in hops.

Kentish Oasts have played their part as a focal point in the development of this very special sector of British agriculture. Today there are now only some 200 hop growers in production, growing just over 3000 hectares of hops, equally split between Kent and the West Midlands. For many, Kent Oasts remain alive, either functioning as hop dryers or many more as dwelling places. They are all monuments to the life and work of the many hop growers through the ages.

This book is the result of 20 years research by Robin and Ivan Walton. For all those with an interest in hops and oasts, be they hop pickers, growers or owners of oasts as dwelling places this book is one of the only definitive works on Kentish Oasts providing a detailed record of their history, the people who worked in them, their equipment and their construction. This book is therefore providing a testament to the Kent hop industry through the ages.

Richard G B Wood
Director, English Hops Ltd

CONTENTS

LIST OF OAST PHOTOGRAPHS

NOTE. Many have now been converted into domestic dwellings, museums, theatres, etc. and some renamed. Others demolished or burnt down.

GENERAL PHOTOGRAPHS

LIST OF FIGURES

Filling the hop pocket

Introduction

DERIVATION OF OAST

The origins of the modern word 'oast' are derived from two old Latin words 'aidis' meaning hearth or house and 'aestus' meaning heat.

Before hops were introduced into Kent in the 16th century, variations on these two basic old Latin words were used to describe any building used to dry materials. Examples of these include 'ast', 'host' and 'nost'.

Flemish merchants introduced words such as 'est', 'east' and 'eest'

Between the 16th and 18th centuries, a kiln used to dry either malt or hops was called 'ost', 'oste', 'oost' or 'oust'. By the 19th century the commonly recognisable word 'OAST' was adopted and referred specifically to a kiln used to dry hops.

HOPS

The primary reason for the building of an oast, is to dry the mature hops and to prepare them to an acceptable standard for the brewer.

The plant can be found growing wild in the hedgerow but normally it is commercially cultivated in 'hop gardens' where the bine is trained up a pole or string, traditionally to a height of between 16 and 20 feet, (6-8ft. is being introduced).

The hop plant, 'Humulus lupulus', is dioecious, its roots are perennial and bine annual, with the male and female flowers growing on separate plants. Wind-borne pollen from the male plant is caught on the brush-like female flower, the burr of which rapidly develops into the mature yellow-green hop cone which is usually ready for picking in the first week of September.

It is actually the golden yellow resin glands (lupulin) which contain the valuable substances required in the brewing of ales, stouts and lager beers. A high Alpha acid content is a requirement that the brewers look for in a good crop.

Over the centuries people have cooked and eaten the fresh new shoots, which they referred to as 'the poor man's' asparagus.

Hops were used by physicians to cure many ailments. Nicholas Culpeper, 1616-54, published a herbal book in which he made reference to the many diseases and ailments for which hops could be used, e.g. "to cleanse the blood, killeth ringworms, cureth yellow jaundice and all manner of scabs".

Freshly Picked Hops

During the 14th century the hop was being cultivated in the Netherlands, particularly in the Flanders area, and it is thought that they were introduced into England by the Flemish weavers who travelled to this country in search of markets to sell their goods, though these hops were not cultivated on a commercial scale in this country until the early 16th century.

The importance of hop cultivation was recognised by the founding of the 'Order of the Hop' (l'Ordre du Houblon) by John the Fearless, Duke of Burgundy, cousin of King Charles VI of France, in the early 15th century.

In 1956, five hundred years after John's death, at the European Hop Growers' Convention held in Strasbourg, members decided to revive this ancient Order and to confer it upon individuals of the member countries who had rendered service to the interests of hop growers in the international field.

THOMAS TUSSER 1515-1580

Tusser, who flourished in the reign of Henry VIII and Elizabeth I, was a gentleman and a practical farmer, and as the result of his labours and knowledge gave the public a whimsical treatise in verse, entitled, 'Five Hundred Points of Good Husbandry'.

He states that the culture of hops was introduced into England from the Netherlands, in 1524; and they are first mentioned in the Statute Book of 1552.

The following verse for the month of August is taken from Thomas' treatise.

"August's Husbandry"

Hops gathering.	If hops do look brownish, then are ye to slow,
	If longer ye suffer these hops for to grow:
	Now sooner ye gather, more profit is found,
	If weather be fair, and dew off a ground.
Increasing of hops.	Not break off, but cut off, from hop the hop-string,
	Leave growing a little, again for to spring:
	Whose hill about pared, and therewith new clad,
	Shall nourish more sets, against March to be had.
The order of hop gathering	Hop hillock discharged of every let,
	See, then, without breaking, each pole yet out get;
	Which being untangled, above in the tops,
	Go carry to such as are plucking the hops.
Hop-manger.	Take soutage, or hair, that covers the kell,
	Set like a manger, and fastened well;
	With poles upon crotches, as high as they breast,
	For saving and riddance, is husbandry best.
Save hop-poles.	Hops had, the hop-poles that are likely, preserve
	From breaking and rotting, again for to serve;
	And plant ye with alders or willows a plot,
	Where yearly, as needeth, more poles may be got.
Drying of hops.	Some skilfully drieth their hops on a kell,
	And some on a soller, oft turning them well.
	Kell dried will abide, foul weather or fair,
	Where drying and lying, in loft do despair.
Keeping of hops.	Some close them up dry in a hogshead or fat,
	Yet canvas or soutage is better than that:
	By drying and lying, they quickly be spilt,
	Thus much have I shewed; do now as thou wilt.

Glossary

Crotches	-	Forked sticks
Kell	-	A kiln
Soller	-	A loft or upper chamber
Soutage	-	The cloth in which hops are generally packed

GROWTH DISTRIBUTION

Hops are grown these days in a climatic zone around the world which includes in the Northern Hemisphere U.S.A., Great Britain, Europe, China and Japan, covering 97% of the total acreage.

The remaining 3 % is in Argentina, South Africa, Australia and New Zealand in the Southern Hemisphere.

In 1992 hops were grown in 29 different countries throughout the world. The single largest hop producing country was Germany, U.S.A. second and England in sixth place.

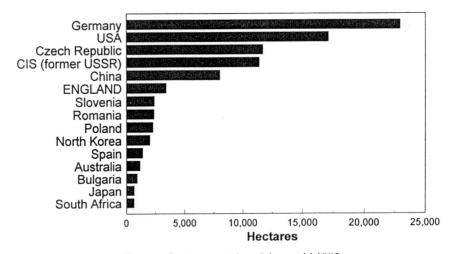

Hop producing countries of the world 1992

However during the peak period of hop acreage (1870-78)in Great Britain hops were grown in a total of 51 counties; 38 in England, 8 in Wales and 5 in Scotland, covering as far north as Aberdeen and west as Carmarthen.

Only the South East and West Midlands in England continued with production after the market declined in the 1890's.

In 1844 Charles Muggeridge produced a map of the hop districts of Kent for the Directors of the London and Dover (South Eastern) Railway Company. The map showed the divisions of growth and the number of acres in cultivation in each parish. It also indicated the two existing and six intended railways. It was obviously drawn to show the commercial potential for transporting Kent hops to the London market.

4

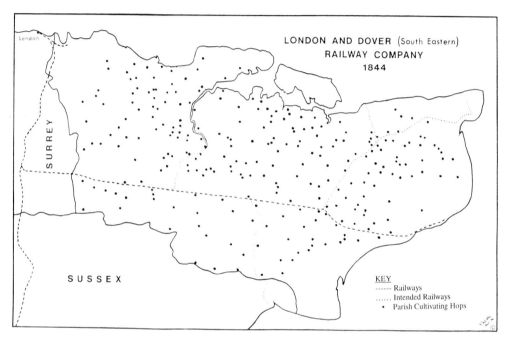

Kent Parishes Cultivating Hops 1844

At the time of mapping hops were still being grown to the west of the county on a line through Bexley, Orpington and Westerham.

By 1878 at the height of production, hop cultivation had encroached closer to London, into the Bromley area.

During our eighteen years of research in Kent we have found no records however of hops being grown nor evidence of oasts on the Isle of Sheppey.

The English industry is now centred in the South East, with about 60% of the area in Kent, Sussex, Surrey and Hampshire hop gardens and the remainder in the hop yards of Herefordshire and Worcestershire in the West Midlands.

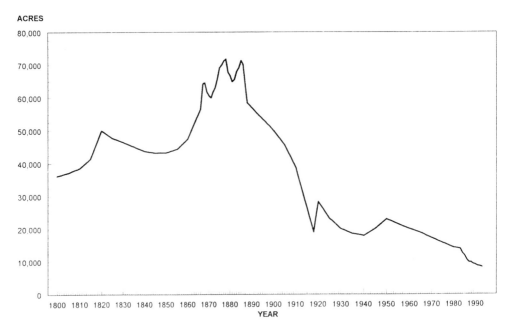

British Hop Production

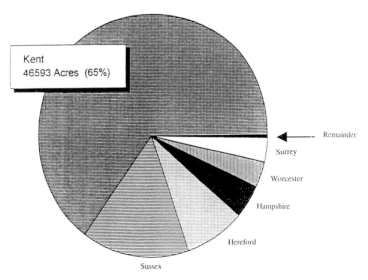

Kent
46593 Acres (65%)

Remainder

Surrey

Worcester

Hampshire

Hereford

Sussex

British Hop Production for the Peak Year: 1878

Daniel Defoe, the famous author of "Robinson Crusoe", also published in the late 1720's "Defoe's Tours" in which he mentions the area around Maidstone as being the first site in England where hops were planted in any appreciable quantity. The largest single area under cultivation at that time is quoted as being 6,000 acres around Canterbury, although he does state that he is unable to vouch for the authenticity of this value.

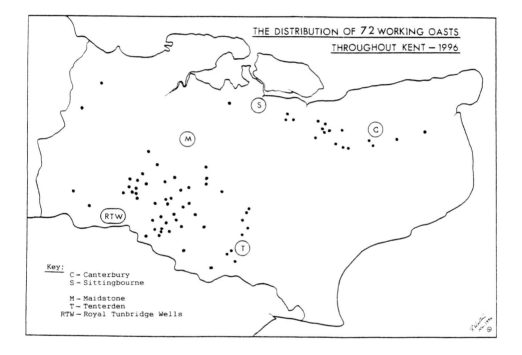

The map shows a number of oasts distributed along the main A2 road between Sittingbourne and Canterbury but by far the greatest concentration of oasts in the county today are located in the Weald roughly within the triangle formed by the towns of Maidstone, Tenterden and Royal Tunbridge Wells.

ROMNEY MARSH

The Romney Marshes are a most unlikely hop farming area, but there is evidence of some successful hop growing/cultivation.

In 'Kellys 1899 Directory' it lists Mr Stephen Hickman of 'Moat Farm', Ivychurch as a Farmer and Hop Grower.

An oast had been built in the village on the area opposite the 'Bell Inn', public house. The pub still stands but the oast site is now a childrens' playing field. The oast stood by the roadside where the field gate is now positioned.

The accompanying unique photograph shows a procession led by tambourine and accordion players in 1963 on their way to the Flower Festival and Sports Day to be held on the ground behind the oast. Part of the village hall can be seen on the right and the church tower above the oast roof.

Ivychurch Oast Circa 1963

ISLE OF THANET

Hops were grown in almost every area of Kent during the massive increase in acreage between 1860 and 1885. The only known exception being the Isle of Sheppy.

There are two oasts still standing in the North Forelands area on the Isle of Thanet.

At East Northdown Farm a large three storey brick oast built in 1886 remains, with a slate covered roof and a single 16ft square kiln. The stowage windows have segmental brick arches and cast iron pivoted sashes. Permission has been given by the Local Planning Authority (in 1994) to convert the oast into four domestic dwellings.

East Northdown Farm

Elmwood Farm has a small single 16ft roundel oast constructed in 1836. After many years of hop drying it was subsequently utilized as a shelter for sheep and donkeys.

In 1995 it was converted into a domestic dwelling and can be seen on farmland to the west of the North Forelands Lighthouse.

A derelict oast at East Peckham

10

OASTS

The earliest book we have discovered describing an English Oast was published in 1574. It is entitled, "A Perfite Platforme of a Hoppe Garden" and the author, a Reynolde Scot, describes his own experience as a hop grower in the parish of Smeeth, near Ashford.

Scot's book is illustrated by numerous woodcuts but unfortunately only two actually depict the oast construction; one the plan of the oast and the other a close-up of the furnace.

The 'oste' was simply a rectangular building 18 feet long by 8 feet wide, with 9 feet high walls, much smaller than later 19th and 20th century oasts. The building was divided into three 'roumes'; the centre portion containing the furnace, the 'forepart used for the green hops and the 'hynderpart' for the dried hops. The furnace or 'keele' was built of bricks, 5-7 feet in length, 2 feet 6 inches high and 13 inches internal width. The upper part of the sides and rear of the furnace walls were 'honeycombed' brickwork allowing the hot air, smoke and fumes to escape. The top of the furnace consisted of two rows of bricks leaning against each other to form a pitched roof.

The drying floor was only 5 feet above the ground floor and consisted of strips of wood 1 inch square, spaced a quarter of an inch apart and supported in the centre by a wooden beam entirely spanning the oast. The hops were laid out on this floor to a depth of 9 inches. It would appear that they were placed directly upon the strips of wood without the use of any form of cloth. The main frame structure of the building was probably of local timber.

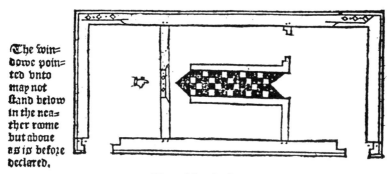

Plan of Scot's Oast

11

A Perfite platforme

of a Hoppe Garden,
and neceſſarie Inſtructions foꝛ the
making and mayntenaunce thereof,
with notes and rules foꝛ refoꝛmation
of all abuſes, commonly practiſed
therein, very neceſſarie and
expedient for all men
to haue, which in any
wiſe haue to doe
with Hops.

Nowe newly corrected and augmented
By Reynolde Scot.

Prouerbs.11.
Who ſo laboureth after goodneſſe, findeth his deſire.

Sapien.7.
Wiſedome is nymbler than all nymble things.
She goeth thorough and attayneth to all things.

❧ Imprinted at London by Henrie
Denham, dwelling in Pater noſter
Rovve, at the Signe of
the Starre.
1576.

Cum priuilegio ad imprimendum ſolum.

Title Page of Second Edition

This form of construction continued well into the 18th century. Richard Bradley's book published in 1729 shows a similar plan.

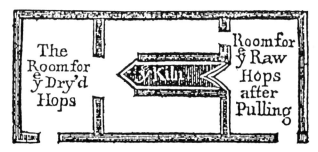

The buildings we now regard as typical Kent oasts with 12-16 feet square kilns and a pyramid roof joined to a rectangular stowage, are a relatively recent introduction. These oasts evolved in the latter part of the 18th century.

In the early part of the 19th century the 12-18 feet circular kiln, with the conical roof, was introduced. It was considered to be a more efficient design than the older square kiln, but this assumption was found to be incorrect and towards the end of the century the design for new oasts built reverted to slightly larger 16-20 feet square kilns.

19th Century Oast - The Roundel was added in the Victorian Era.

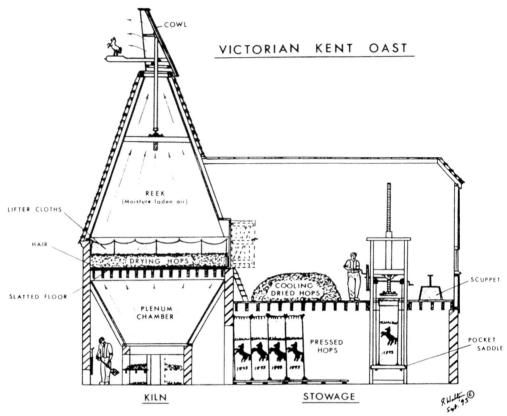

Victorian Kent Oast

MODERN OASTS

Gone are the days of constructing roundels for a new oast. Nowadays they consist of a single story steel or concrete portal framed building, clad with corrugated sheeting under a pitched roof. Modern hop farming has necessarily also incorporated a number of mechanical techniques. The dimensions will vary depending upon whether the hop picking machines are housed within the same building.

In the hop gardens the bines are mechanically gathered using 'bine pullers' on a fleet of tractor/trailers. An arm from the bine puller, itself mounted on the front of a trailer, deflects the hanging bine on to a cutter, severing it from the roots. The loose end is wrapped around a horizontally-mounted rubber wheel, which pulls the bine off the overhead wirework onto the floor of the trailer.

14

On arrival at the hop-picking machine, the bine is hooked onto an endless chain which conveys the suspended bines into the picking machine, where metal loops ('fingers') protruding from a steel drum pluck off both hops and leaves. The bare bines, on leaving the picking machine, are chopped into short lengths and discarded on to a heap outside the building to decompose! The majority of the leaves are blown away by fans and any remaining leaves are separated from the hops by a series of elevated endless belts.

The Continuous or Progressive bin drying system was first introduced in 1963, but over the years has been uprated and is now widely used. The hops are fed from the picking machine along a conveyor and discharged evenly into steel bins, which are 3 feet deep and either 8 feet square or 15 feet long by 6 feet wide, having an open top and mesh base.

A bin full of green hops is added to a line of bins already containing hops in various stages of drying.

Up to 24 bins mounted on rollers can span large underground plenum chambers. Hot air from oil burners is fed by fans along ducts from an adjacent area into the chambers. The drying air percolates up through the mesh base of the bins and out of the open top. In the early stages the resulting moisture laden air ('reek') is drawn up by fans mounted in the roof space and then expelled through domed ventilators in the roof.

Electronic sensors continually monitor the progress of each bin. As the end bin of cured hops is removed and the contents sent for pressing, a fresh bin of green hops is added at the beginning of the line.

Finally, a hydraulic operated press produces rectangular bales weighing approximately 200lbs. These bales are dispatched to a warehouse, brewery or alternatively to a processing plant, where the hops are formulated into pellets or a liquid extract.

Typical Modern Hop Drying Buildings.
(Lily Hoo, Beltring)

Chapter 1

Construction and Materials

FOUNDATIONS

The foundations to the early oasts consisted mainly of dwarf ragstone walls on which the timber framed buildings were constructed.

A small number of oasts were built during the 18th and 19th centuries without adequate foundations. Some have their walls only about 18 inches below the ground level and subsequent ground movement by the action of mostly frost and drought has caused cracks in the superstructure. Brick or concrete buttresses have been constructed in many cases to prevent collapse of the oast, also iron straps (roundels) and tie irons (stowage's and square kilns) have been used to support the structure. When converting these buildings to domestic dwellings it is often necessary to also underpin with concrete.

Most of the large oasts of the 19th century have stepped brick foundations (see 'Nashenden Oast' on page 69) also brick plinth walls up to 3 or 4 courses above ground level which reduce to the general 9 inch wall thickness, on a concrete strip foundation.

To prevent damp rising up the walls a small number of oasts (from the 19th century) have a damp proof course built in. This damp proof course (DPC) was either a layer of hot bitumen spread over the brickwork or 2 courses of roofing slate laid in a breaking joint pattern bonded in cement mortar.

Oasts being converted into domestic dwellings must have a DPC to conform to the current Building Regulations. One common method is to inject a chemical into the wall to form an impervious water barrier at a height of not less than 6 inches above the ground level.

A very expensive osmotic DPC incorporating copper strips let into the walls, has been used on a conversion in Marden.

TIMBER FRAMED STOWAGE

Prior to the 18th century, the majority of oasts were constructed entirely of softwood or reclaimed oak timber. The frames would have been prefabricated in sections on the ground, with the joints carefully fitted and Roman numerals cut into the timber at the joints to aid reconstruction. Externally the frames were clad with tarred horizontal boards. The internal surface of the studding was covered with 'Wattle and Daub'. Wattle is the name given to a closely constructed hurdling of pliable material, either hazel or ash sticks, or riven oak laths. Daub is an adhesive mixture of clay, chopped straw or cow-hair and cow dung, spread over the wattle to give a flat surface to the wall.

During the period 1784-1850 the introduction of the brick tax resulted in a reduced usage of bricks, particularly for the stowage walls where timber was often favoured. The ground floor walls were of brick, with a timber framed weather boarded upper storey. The majority of upper storeys were covered on the inside by softwood tongued and grooved boarding. The outside surface was covered with either softwood feather-edge weather boarding (tarred or painted white) or softwood battens nailed horizontally to the studding, with handmade red clay tiles hung from, or nailed to, these battens.

With the oak framed buildings, brick noggings were used to fill the space between studs in the framework and were usually laid diagonally or herring-bone fashion, rather than the normal horizontal layers. In the early 20th century corrugated iron sheets were sometimes used to clad the stowage, as a quick and cheap alternative.

PORTAL FRAMED BUILDINGS

Modern oasts constructed in more recent years have generally been built as prefabricated industrial type structures, rather than to the traditional design. This gives more clear space in which to operate and is quicker and cheaper to erect.

Reinforced concrete frames were first used, clad with corrugated asbestos sheets or brick walls under an asbestos roof. This type of building is also often used to house the hop-picking machinery. Mild steel framed construction followed, which are normally clad with either corrugated asbestos or aluminium sheets.

Most of the reinforced concrete portal framed buildings erected in Kent, would have been cast in steel moulds in the 1950's and 60's at the 'ATCOST' workshops in Paddock Wood.

This photograph at Yalding depicts both types of construction on a single site. The completed concrete framed building in the background houses the picking machine and the unfinished steel framed structure in the foreground will eventually contain the drying and pressing equipment.

WALLS

Brick

In 1784 the Government imposed a brick tax and this lead to many oasts being constructed of alternative materials. This tax was abolished in 1850 and the majority of oast walls, both kiln and stowage, since this date have been constructed of 9 inch thick brickwork.

The bricks of the 18th century were hand-made from local clay, laid in lime mortar to English bond i.e. the bricks laid in alternate courses (rows) of headers and stretchers or English Garden Wall bond, which has three rows of stretchers and one row of headers. The shrinkage and distortion in bricks, caused during firing, made it necessary to lay the bricks with wide mortar joints, often strengthened with flint chips or small stones, a practice known as 'galleting'.

Oasts built in the 19th century were mostly red or multi-colour hand-made bricks laid in lime mortar to Flemish bond i.e. each course consists of alternate headers and stretchers or Flemish Garden Wall bond having one header followed by three stretchers in each course.

Although English bond is the stronger, it was probably because Flemish bond was considered the more attractive arrangement that it became the most popular style of bricklaying.

The 20th century brick-built oasts are constructed from machine-made clay or sand/lime bricks, to a British Standard size, laid in cement mortar to Flemish bond.

In the modern steel or concrete framed stowage the brick infill panels would be 4½ inch thick laid stretcher bond or 9 inch Flemish bond.

Alternatively, the frame can be clad with corrugated asbestos cement or aluminium sheets.

Yalding

BRICKWORK BONDING

Key: H – Header (end of brick)
    S – Stretcher (face of brick)

STRETCHER

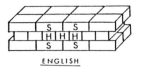

ENGLISH

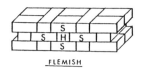

FLEMISH

ARCHES

WELSH

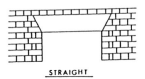

STRAIGHT

SEGMENTAL

RELIEVING ARCH

QUOIN (corner)

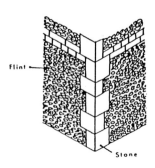

Flint

Stone

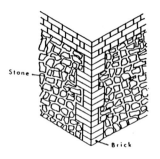

Stone

Brick

20

Flint

A few walls were built of 'knapped' flint, i.e. a flint found in the local chalk of the North Downs, broken to expose the hard shiny inner surface.

These walls were liable to be deficient in longitudinal and transverse strength because of the small size of this material. To overcome this weakness, lacing courses, of two or three brick rows, were built in at intervals of approximately 3 feet in height, normally corresponding to floor and window levels. Cylindrical type kilns - roundels, having no supporting quoins (corners) required similar vertical lacing courses to tie the structure together.

Quoins (corners) of knapped flint, random ragstone walls, and jambs (corners) to windows and door openings, were generally built with blocks of rough or quarry face stone, or alternatively block brickwork, giving strength and also straight lines.

Sandstone

Another local material that has also been used in wall construction is sandstone. This deep yellow coloured quarried stone is cut into random size blocks which then require an experienced craftsman to lay them.

Ironstone

Ironstone is a dark brown coloured sedimentary rock that was quarried in the Weald of Kent up until the early 19th century.

A seam of ironstone was worked in the Borough Green area and many buildings, including single and double roundel oasts built of ironstone can still be found at Comp, near Platt.

Also wood from the Wealden forests was utilized to fire the charcoal furnaces used to extract the iron from the rock.

Kentish Ragstone

The main alternative material used for the kiln and stowage was Kentish Ragstone, a local green/grey sedimentary rock.

It was necessary to increase the wall thickness to approximately 16 inches because of the random size of this material, to give the required stability.

Block cut ragstone was rarely used because it was too expensive for this type of building.

PILLAR-BOX HOUSE OAST

A delightful picturesque oast can be found in the West Kent village of Heaverham, just off the Pilgrim's Way.

The typical stowage is built of 9 inch brick laid Flemish bond. But the twin roundels are interesting in that their bases are constructed of 15 inch thick 'galletted' ragstone (to a height of 9 feet) and then a reduction to 9 inch thick brick up to the eaves. The reason for this is not purely cosmetic as the uniformity of this type of brickwork enables ease of construction of the drying floor and roof eaves.

Ragstone and Brick

Great Portland, Knockholt,Nr Sevenoaks
Knapped Flint Wall

Chiddingstone Oast

Timber framed stowage with ground floor brick infill panels and vertically hung upper storey tiles.

23

SCARLETTS OAST, COWDEN

At 'Scarletts Farm' on the borders of Kent, Sussex and Surrey, there is a listed oast, consisting of a single 18 feet diameter kiln partly constructed of a local quarried stone.

Blocks of stone are laid in horizontal courses, the height varies from 12 inches at the base down to 6 inches at the top course. Some of the base blocks are 24 inches in width getting smaller in the upper courses but still retaining a bond. All blocks have been dressed to the curvature of the kiln.

Just below the drying floor level, red bricks laid header bond continue up to the roof eaves. The roof is covered with tiles.

The original stowage has been demolished, with only the foundations and part of the wall abutting the kiln entrance doorway remaining.

Local Quarried Stone

STREET FARM OAST, BOXLEY

Originally this substantial building built in the last century, consisted of a 50ft by 20ft stowage, with king post roof trusses and two 16ft square kilns all under pitched tiled roofs.

Three of the stowage walls are constructed of 13½ inch brickwork, with red bricks at the front and multicoloured stocks to the rear. Unusually the southern flank wall consists of ragstone and flint to a height of 4ft and the remains in 16 inch chalk blocks.

When the Rochester Bridge Trust initially purchased the farm in 1950, the oast roof was in a very poor state of repair. Subsequently the kiln roofs were demolished and replaced with corrugated asbestos cement sheets. The stowage roof was stripped of tiles and similarly covered with sheets over the original wooden roof trusses.

In 1987 this Grade II listed oast was sold for domestic development. The District Planning Authority granted permission to the developer in 1988 to convert the oast into two 4 bedroom units, subject to the condition that the cowls would be reinstated in order to preserve the visual character of a typical Kentish Oast.

Chalk Blocks

25

TILDEN OAST

Within the Wealden clay there are several thin bands of limestone, formerly worked as Marble, which are made from the shells of a fresh-water snail e.g. Paludina.

A fine example of this material in the form of an 8 feet high monument can be seen in the centre of Bethersden, commemorating in 1935 the 25th year of the reign of His Majesty King George V.

In the Headcorn area there is a unique oast which has one of its two roundels built of Bethersden Marble, which is actually pale brownish grey in colour.

The oast has been converted into a domestic dwelling with the wooden hop press remaining within the building, as a reminder of the buildings former use.

Bethersden Marble

ARCHES AND LINTELS

Arches to doors, windows and internal openings in walls are mainly of two courses of brick, either segmental or semi-circular in shape. But there are many exceptions.

Unusual examples can be seen at the following sites

(i)   Ightham   - a three-storey ragstone oast where the window openings are triangular shaped at the head.

(ii) Pluckley   - elliptical ragstone arches to door openings and elliptical relieving arches above windows.

(iii) Yalding   - brick-built stowage with Gothic style window arch.

The small observation windows and air vents built into the walls of some kilns have 'Welsh' arches i.e. dovetailed brick keyed into the general brickwork.

The very small narrow 'slit-like' open air vents at the base of the kiln are bridged by a brick.

Timber lintels are commonly used to support the inner skin of walls above doors and windows.

The timber 'turning-piece' used in the construction of segmented brick arches is normally left in position above the door and window frames, as at Broadwater Farm.

Broadwater Farm, East Malling

27

Ightham Oast

Pluckley Oast

28

EAVES

Construction of the kiln wall at the eaves consists of 'oversailing' courses of brickwork or dressed ragstone. This supports the lower tiles or slates and enables the roof to throw the rainwater clear of the building, since there are no gutters for collection and discharge of rainwater. A variety of 'oversailing' brickwork techniques are used to give strength and to add a decorative feature to the building.

The stowage eave construction is identical to that used in a domestic house i.e. fascia and soffit boards but not necessarily with guttering.

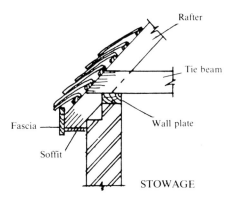

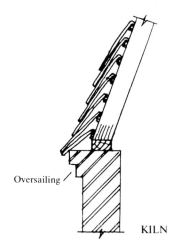

Details of Eaves

STOWAGE: ROOF CONSTRUCTION

Roof construction is mainly dependent upon the span of the building and the designs have changed through the centuries.

The 17th century oasts had roof members of hewn oak, whereas in Victorian times, softwoods, such as Douglas Fir and Scots Pine, replaced the oak because they were cheaper and lighter in weight.

SINGLE AND DOUBLE ROOFS

The early oasts for example, 'Reynolde Scots' with an 8 feet span, only required a single roof comprising solely of rafters and a tie beam. When this tie beam was positioned high up in the roof, it was referred to as a collar.

As the size of stowage increased a double roof construction was introduced. A horizontal member, called a purlin, was incorporated to provide intermediate support, thus permitting the use of comparatively smaller rafters.

TRIPLE OR FRAMED ROOFS (TRUSSES)

On the large roofs the combined weight of construction materials was supported by a number of separate frames or trusses which were placed at 10-12 feet centres.

With spans of up to approximately 30 feet a King Post Roof Truss with a single vertical member was used. It was uneconomical to construct this type of truss on spans of greater than 30 feet and so the Queen Post Roof Truss with two vertical members was introduced.

Joints

Ridge boards were not introduced until the 18th century and therefore it is possible to date oasts of the 16th and 17th centuries by their absence. Additionally the rafters were either bridle jointed or half-lapped together at the apex and held by a wooden dowel, with the Queen Strut members morticed and tenoned together and held by wooden pins.

In Victorian times the truss members were tied-in by either iron bolts or straps. Modern roofs use only steel nails to secure members.

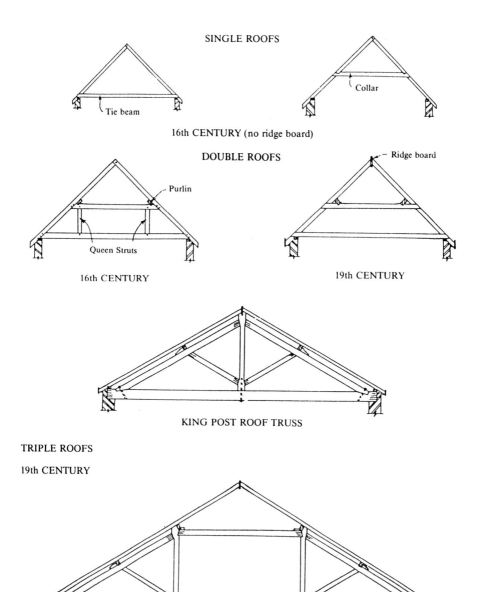

SINGLE ROOFS

Tie beam

Collar

16th CENTURY (no ridge board)

DOUBLE ROOFS

Ridge board

Purlin

Queen Struts

16th CENTURY

19th CENTURY

KING POST ROOF TRUSS

TRIPLE ROOFS

19th CENTURY

QUEEN POST ROOF TRUSS

R Walton

Types of Roof Construction

31

STEEL ROOF TRUSSES

Some of the oasts built during the 1930's had the roof covering supported by steel roof trusses fabricated from lengths of mild steel angle, bolted together at the intersections.

Bell No. 5 oast at Beltring (built in 1939) used this type of roof construction.

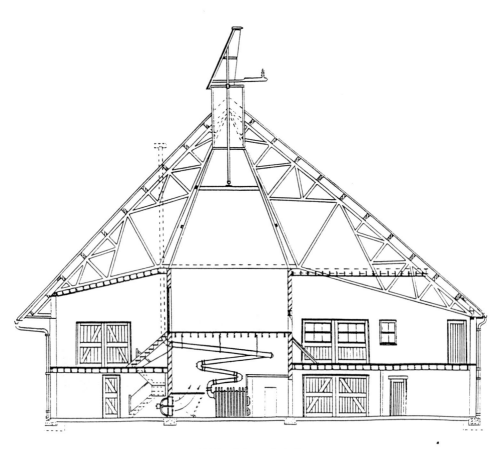

Bell No. 5 Oast, Beltring
Permission to reproduce this cross section drawing was kindly given by
Nicholas Redman, Archivist, Whitbread plc.

ROOF PROJECTION

The hop press ram in some oasts has to be situated close to an outer wall to give maximum use of the floor area for cooling.

To enable the ram to be raised to its full height, projecting between 12-18 inches through the sloping stowage roof, a small 6 inch square wooden box-like structure is constructed for its projection.

Tutsham, West Farleigh

33

STOWAGE: ROOF COVERINGS

Kent Peg Tiles

Since the 16th century the standard 10″ x 6½ ″ x ½″ clay tiles have been laid on laths of riven (split) oak or softwood and held by oak or softwood pegs driven through the holes in the tile and hooked over the lath.

These tiles were made by hand and because, like bricks, they were thoroughly baked were liable to distortion.

Some tilers bedded the tiles in lime mortar whilst others bedded them on a very thin layer of hay or straw sometimes mixed with mud to weather-proof the roof.

In the early 20th century galvanised nails, with large diameter shanks and heads, replaced the wooden pegs. Where roofs are now being stripped for domestic buildings the galvanised nails are being replaced by either alloy or plastic ones. As the galvanization does degrade, these modern replacements are more durable and help avoid the problems of rusty nails expanding to eventually crack the tiles.

They cannot be hung on roof pitches of less than 40° because of the size and curvature of the tiles.

Slates

Slate is quarried from an argillaceous rock, i.e. one originally deposited as a clay or very fine mud, which has been subjected to the action of great heat and pressure over a long period of time. Since 1820 the main source of supply has been from Wales.

Slates are cut into various standard sizes and are referred to under a series of names: Princess, Duchess, Marchioness, Countess, Viscountess and Ladies, in decreasing order of size.

The method of fixing is by two large flat-headed galvanised nails, driven through punched holes, into 2 inch by 1 inch softwood battens, each slate being overlapped at the head for weathering.

Slates can be used on very low pitched roofs, with 22° the minimum.

When exposed to the elements, the life span of the ordinary ¼ inch thick slate is approximately 100 years, after whch they begin to flake and rapidly deteriorate.

KILN: ROOF CONSTRUCTION

Square

The construction of square kiln roofs is similar to that of a hipped roof domestic building.

Common rafters are fixed to the wall plate and cowl kerb at 16 inch centres. These can either be in one length, supported at the mid point by a purlin or in two lengths overlapping each other at the purlin.

A **hip rafter** supports each corner with the horizontal purlin nailed at the centre span. At each corner shorter **jack rafters** are fixed between the wall plate and the side of hip rafter.

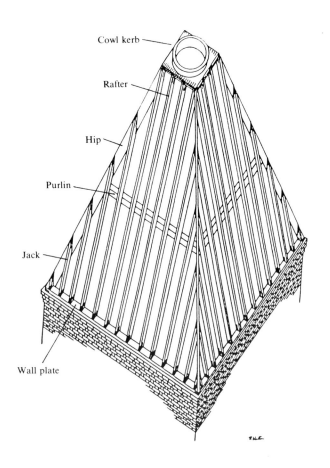

Roundel

The roof to a roundel is normally constructed in one of two variations. Small diameter roofs have rafters running from the wall plate right up to the cowl kerb but this method causes a congestion of members at the narrow top. The alternative, used mainly in the construction of large diameter roofs, is to have four or six principal rafters spanning the wall plate to the cowl kerb. Then an intermediate timber 'ring' is constructed and jointed to the principal rafter by mortise and tenon joints. Additional rafters from the wall plate then terminate at this ring with a lesser number of rafters fixed from the ring to the cowl kerb.

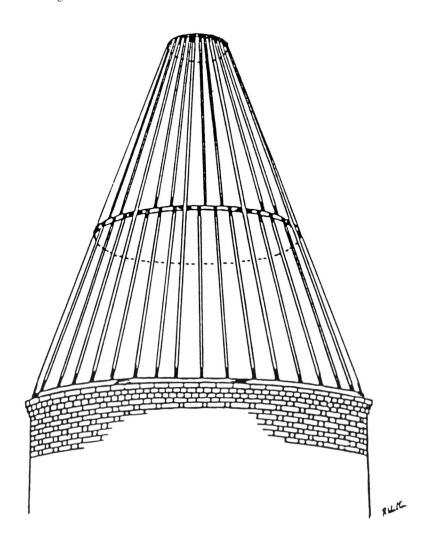

KILN: ROOF COVERINGS

Square

Both standard tiles and slates have been used for the square kiln, with its pyramid-shaped roof. A large amount of cutting is necessary for the bonding at the hips (corners). Normally with tiles a special hip tile is bedded in mortar over the corner intersection. But with slates, either a hip tile or more commonly a strip of lead (flashing) is dressed over the intersecting slates.

Roundel

The roundel, with its conical roof, requires special tiles. These taper in length, with two holes at the narrow end for the pegs or nails. The slater, however, had to individually cut and hole each slate to conform to the necessary taper.

Eaves to Kiln

To enable the tiles or slates to oversail the very thick walls of a ragstone-built roundel, sprocket pieces or rafters are nailed to the foot of the ordinary rafters. When this technique is used it gives the roof a 'bell-shaped' appearance.

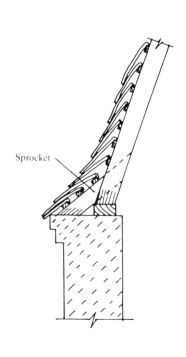

Sprocketed Eaves

Oast Manor, Gillingham

37

BRICK ROOF TO ROUNDEL

A technique which called for considerable skill was to build the lofty cone of 4½ inch brickwork with 9 inch internal piers, up to the cowl support beams. The exterior was cement rendered and then coated with a mixture of tar and pitch, applied while hot.

This type of roundel construction required neither internal plastering or external tiles/slates and therefore potentially had the lowest maintenance cost.

Brick Roof to Square Kiln

This rare type of construction was used to build the three square kiln roofs at 'Biggenden Oast', Paddock Wood, in the late 19th century.

Bill Davis of Five Oak Green, started work on this farm at the age of 13 years in 1914 and retired after 52 years of service, of which 47 years were spent in drying hops, firstly using coal fires and during the latter years a 'Urquhart' oil burner

Biggenden Oast, Paddock Wood

A second oast in the Paddock Wood area at Whetsted has four brick roof kilns of which two are square. This oast is unique, it also incorporates vertical sliding sash windows. These elaborately constructed and expensive type of window would have initially been made for a Victorian domestic dwelling and later utilised in this oast. Both oasts have now been converted into domestic dwellings.

Cooklands Oast 1905

KILN VENTILATION

The early rectangular oasts used a very simple form of roof ducting to allow the moist air from the drying hops to escape naturally. During the Victorian era the most numerous form of ventilators were the elaborate and prestigious cowl, which dominated the landscape. The majority of the oasts built or converted during the early part of the 20th century used the wooden louvred system because of cheapness and ease of construction. These were built either along the ridge of the rectangular oast or on top of the square kilns.

Within the last thirty years a further improvement has been found necessary to cope with the efficient high volume circulating fans. The cowls and louvres have been replaced with large hinged wooden flaps.

Louvred Ventilation (Selling)

Hinged Shutters (Wateringbury)

THE "WATSON" VENTILATOR

From the very early days hops were laid to a depth of 9 inches on the drying floor, above a coal or charcoal fire. The air passing through the drying hops could easily escape via the cowl.

After the 1939-45 war, with the drive for improved efficiency the depth of drying hops was increased to three feet or more. The larger volume of air required to dry them in the same time caused problems for the conventional cowl which often could not cope and a back pressure formed.

To alleviate this problem many farmers constructed side ventilators in the roof of the kiln.

In the 1960's a Mr L. H. Watson consultant engineer, from Swanley, designed and installed an improved ventilation system along the roof ridge (or at the apex) in many Kent oasts. This comprised of a wooden pitched roof structure, with hinged flaps on each side, which although unsightly in some cases, proved very effective.

Five Oak Green

COWL

The traditional cowl may be described as an inclined cone, open on one side for approximately one third of its circumference.

The purpose of this wooden construction is purely for ventilation of the kiln, enabling the heavily moisture laden air known as the 'reek' to be drawn from the drying hops below.

The vane acts as a rudder so as to always present the boarded surface to the wind.

The cowl is attached to a central post and the combined weight supported on two beams which are set at right angles to each other and mortised and tenoned into the main rafters at a point approximately two thirds the height of the kiln.

To permit the cowl to rotate freely a metal spigot is driven into the end of the wooden support post (spindle) with a wrought iron sleeve fitted to prevent splitting. The projecting end of the spigot is seated in a recess of an iron plate fixed to the top edge of the support beam. The spindle is kept vertical by a metal collar often called a 'yoke' or 'spider' and has two or three metal stays fixed to the elm weathering ring (curb) of the kiln roof. If the post is square in section metal strips are fixed to wooden packing pieces attached to the post to prevent undue wear as the cowl revolves.

It will be noticed that the cowl lies at a different angle to that of the kiln roof. This is because the backboard or spine of the cowl is deliberately pitched at a lower angle than the roof so that the top end of the spindle can be tenoned into the spine to give maximum strength of construction.

A tie member is bolted at right angles to the spindle and attached to the lower semi-circular rib to give added rigidity.

Alternatively, the rib can be completely circular thus requiring no tie member. The boards forming the cowl overlap each other from the spine towards the opening and are held with copper nails to the wooden ribs.

Two parallel boards set on either side of the opening, resembling blinkers, see sketch, can often be seen in the Kentish Weald and assist in setting the cowl to the wind. The top end of the cowl boards are nailed to a small wooden block and are covered by a cap which weathers the exposed timbers. The vane is known locally by many different names, common examples are: flyboard, finger, etc. It passes through a mortice hole in the spindle and then into the spine or backboard. A wooden wedge is driven through the vane, behind the post, to secure it. This also gives additional strength and stability to the overall cowl construction.

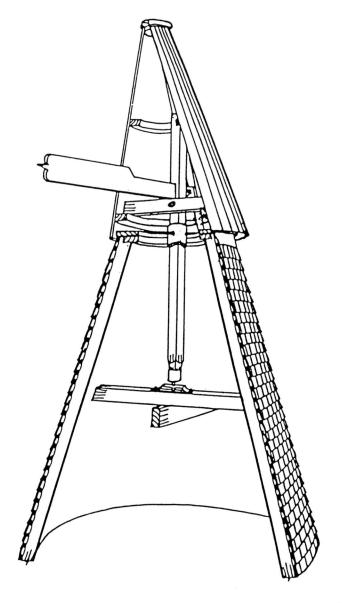

Cowl and Section Through Kiln Roof Showing its Support

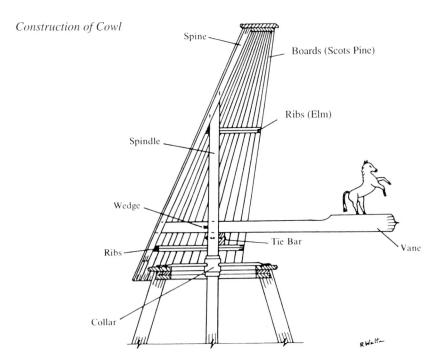

Spine

Boards (Scots Pine)

Ribs (Elm)

Spindle

Wedge

Tie Bar

Vane

Ribs

Collar

R Walt—

Many farmers take the opportunity to display their personal interest on the vane design. The rampant horse of Kent is very common and other unusual and interesting figures include a complete hunting scene, a pig and a tractor.

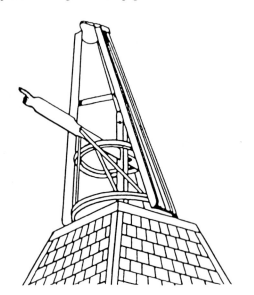

COWLS

Witch -Moon Silhouette

HUNTING SCENE
Tatlingbury Oast, Capel, Nr Tonbridge

47

47

Removal and Repairs

When repairs are necessary to the cowl this cannot normally be carried out with the cowl in position because of the difficulty of access.

To enable the village wheelwright or carpenter to carry out any work, a team of three to five men is required to carefully lower the shell to ground level. Alternatively, a mobile crane is sometimes employed especially where a large number of cowls are to be removed.

The height of these cowls can vary between 5 and 15 feet with most being about 10 feet.

The actual procedure for lowering is as follows:

Firstly a long stout wooden pole is erected, up through the roof opening and projecting out of the cowl. A wooden pulley block is attached to the top of this pole and a rope passes through the block and is anchored to a screw eye or bolt fixed to the spine. The coach screw, securing the backboard to the spindle, is removed and the wedge holding the vane in position is knocked out and the vane is withdrawn.

The cowl is now free to be lifted off its spindle and lowered carefully to the ground, whilst the spindle is lowered down inside the kiln. To maintain stability and prevent any damage to the kiln roof whilst lowering the cowl, two ropes are attached to the bottom ring or tie bar and control is ensured by two men at ground level.

Traditionally the cowls are painted white, usually with up to five coats applied. To reduce the frequency of this costly operation, a technique currently used is to spray the painted surface with glass fibre to improve durability.

At Chart Court Oast, Little Chart near Ashford, the cowls of the four kiln oast are painted red and green in the racing colours of Chester Beatty. He owned racing stables across the valley which could be seen from the oast.

Another exception is the small black cowl on Barn Oast, Biddenden.

In fact, not all cowls are conical in shape. An example of an octagonal shape can be seen on page 79 but the procedure remains the same.

REMOVAL OF COWL
Bearsted Green Oast

Replacing Cowl at Offham

49

FOUNTAIN WORKS, EAST PECKHAM

The construction and repair of oast cowls is a highly specialised task, which calls for a very high standard of technical knowledge and craftsmanship.

A family firm of builders at East Peckham who specialise in this work, still use the traditional method for the removal and replacement of cowls; a pulley block lashed to the top of a wooden pole, which is raised up through the kiln roof.

The firm is currently managed by David Holmes, who can be seen to the right in the accompanying picture. He can also be seen hoisting the repaired cowl onto the kiln roof on page 49. His father-in-law, Arthur Brooks, was a wheelwright and village carpenter, is standing next to the cowl. The example shown here measures about 6 feet 3 inches in height but the tallest one erected so far has measured 12½ feet in height.

During the early summer of 1977 this firm removed, refurbished and replaced the twenty cowls on the four oasts at the Whitbread Hop Farm, Beltring, and in 1984 constructed the three cowls for the oast at the Museum of Kent Life, Cobtree, Nr Maidstone.

50

HOIST FOR GREEN HOPS

The basic principle on which the three-storey oast functions is to store the freshly picked 'green' hops in the roof space above the cooling floor and continually feed them down through the drying and bagging stages to emerge at the bottom ready for transport to market.

To initially raise the bagged hops to the top storey a hoist, of basically three different designs, is used.

The most often used device consists of a single cast iron pulley wheel secured to the end of a projecting wooden beam. This 'Gin' wheel is normally housed in a 'Lucarne' or 'Bonnet gable' above the eaves. One end of the pulley rope is attached to the bag of green hops, known as a poke, and the loose end tethered to the driving horse, which would be unhitched from the wagon. The horse was led several yards along the road, thereby hoisting the poke up to the top storey.

The second design is a refinement on the 'Gin' wheel idea to give added manoeuvrability with the pulley wheel mounted at one corner of a triangular bracket. This 'gallows' bracket can be of cast iron or wooden construction and is pivoted on two brackets secured to the stowage wall.

A large manually operated winch is the other type sometimes found. It consists of a main wooden spoked wheel of 4 feet diameter turned by an endless rope. This in turn drives a smaller 1 foot diameter wooden drum that is attached to the wheel. The load rope coils around the drum and passes over a roller fixed above the door. These large constructions are always mounted in a purpose made wooden frame in the roof space. A smaller version can sometimes be found mounted within a 'Bonnet gable'.

Mechanically Operated Hoist at Clockhouse Farm, Hunton (1980)

Hoisting Pokes of Green Hops

Winch at Swarling Oast, Petham

DOORS

In the early oasts the design of doors was of two basic types generally depending upon the size of opening. Where this opening was small, which in fact was the majority of cases, a simple and cheap construction consisting of wide vertical boards secured by several horizontal ledges was used. The large doorways required a more rigid door construction and therefore had diagonal bracing members. Both of these types of door were normally hung on wrought iron 'T' hinges, known also as a cross garnet. The internal communication door was usually secured by a wooden button or latch. This oak latch was fixed to the outer face of the door and to work it from the inside a string attached to the latch arm was fed through a small hole in the door. On some doors an additional means of opening from the inside was provided by a large finger hole enabling the latch to be lifted manually.

The outer doors were fitted with a more secure stock or dead lock, consisting of an outer wooden casing containing the metal working parts.

The large oasts of the mid 19th and early 20th centuries had ledged, braced and battened internal doors, with the external door having extra bracing and mounted in a heavy wooden frame. The outer kiln door was normally of this heavy construction but smaller in size and often had louvres in the lower half to assist in ventilation. These doors were hung on heavy wrought iron strap hinges or suspended from metal rollers, so as to slide in front of, or behind, the external wall.

The internal doors were mostly secured by metal thumb latches whilst the external doors were secured by heavy metal hasp and staple with padlocks.

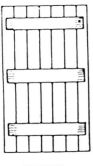

LEDGED

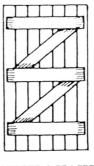

LEDGED & BRACED

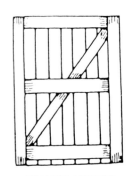

FRAMED, LEDGED
& BRACED

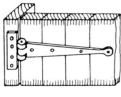

TEE HINGE

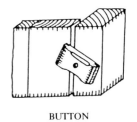

BUTTON

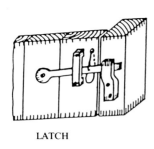

LATCH

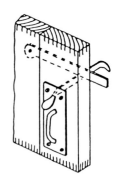

THUMB LATCH

DEAD LOCK

Door Construction and Fittings

WINDOWS

It is noticeable that the design and construction of the windows have undergone several distinct changes over the centuries.

The plan of Scot's oast shown on page 11 indicates that simple vertical square sectioned timber battens, set edge on, were the only means of allowing daylight into the oast. During cold and wet weather, sacking was draped across the openings. Later better protection was afforded by using wooden shutters.

The next advancement was the introduction of horizontal pivoted wooden louvres which were controlled by a simple vertical member, thus giving improved lighting control and security.

Cast iron window frames were introduced in the 17th century, some of which had pivoted or hinged opening sashes. It was necessary to incorporate both vertical and horizontal bars to facilitate the use of small panes of poor quality glass often containing the 'bulls eye' which resulted from the deformed part of the spun glass attached to the glass blower's tube. These cast iron windows were either bedded directly into the brickwork or first set in a wooden frame and then built into the wall.

The Georgian period saw the introduction of wooden casement windows initially with small panes, but as glass became cheaper with improved casting techniques, larger panes became popular and 'bulls eye' pieces disappeared.

Victorian domestic dwelling houses had vertical sliding sash windows but generally these were far too elaborately constructed and expensive for use in a standard oast, but one such example exists at 'Cooklands Oast', Whetsted, built in 1905. Horizontal sliding sashes called 'Yorkshire lights' were used in the construction of a large eight-kiln brick-built oast at East Peckham. An example of a pivoted sash can be seen in an oast at East Northdown Farm, Isle of Thanet.

Since the early part of this century, steel casement windows have been used in several of the new oasts and also to replace decayed wooden frames in old oasts.

The majority of windows are square or rectangular in shape but occasionally an unusual design can be seen. At Bowhill Farm, Yalding, the stowage windows were built to resemble the Gothic style.

Gothic Style Window

55

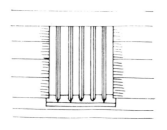

TIMBER BATTENS
(Mullions)

CAST IRON FRAME

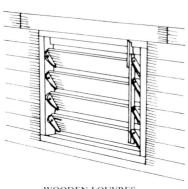

WOODEN LOUVRES

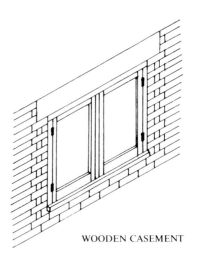

WOODEN CASEMENT

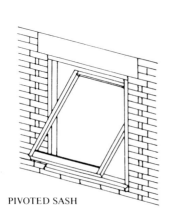

PIVOTED SASH

HORIZONTAL SLIDING SASH

Window Designs

STAIRS

Access to the upper storey in most two-storey oasts is by a straight flight of open tread stairs supported by the outside wall of the stowage with a timber handrail for protection.

Often entry to the upper floor is by a landing, either cantilevered from the wall or supported by posts, known as newels, from the ground level.

Three-storey oasts have the stairs attached to the inside wall and consist of a straight flight of riserless stairs between floors with a length of rope or timber for support.

Sometimes the flight is constructed into a corner of the stowage and consists of treads and risers, of which the lower treads are tapered so as to give the necessary width of tread on the centre line.

Chapter 2

Variations in Design 1745-1993

Although by the 18th Century most of the Public Buildings and Private dwellings
were essentially brick construction, the vast majority of farm outbuildings were still
being built of timber. This began to change after the lifting of the Government Brick
Tax in 1850.

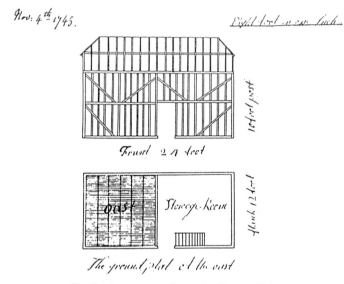

Draft of an Oast at Sutton Vallence (Sic)

Part of Sir Edward Filmer's Promisery Note to Josephus Clout. Nov. 4th 1745:

"And in case the said Josephus Clout should plant hops upon any part of the said
farm and they should come to perfection so as to require an Oast then I do promise to
build an Oast house according to the Draft above upon some convenient part of the
premises".

....

The oast is a rectangular timber-framed building, with a pitched roof, partly hipped
ends and two stories high. It is divided into the drying area and the stowage room for
the dried hops. There are no ventilators, louvres or cowls, indicated.

LITTLE GOLFORD, CRANBROOK: BUILT 1750

A 21 feet by 15 feet timber-framed building with a small lean-to roof at the rear.

The walls are covered with weatherboarding and the hipped roof is tiled, with a cowl built into the ridge.

A timber studded partition divides the oast into two 10 feet 6 inch compartments. One half was sub-divided by a dwarf wall, with two coal fired brick furnaces in the front part and ladder access to the drying floor at the rear. The other half was used as the cooling floor.

The rear door opening has recently been enlarged and the lean-to roof shortened, to allow a car to be garaged, where once the ladder and two furnaces stood.

LITTLE BEWL BRIDGE FARM, LAMBERHURST

This very old oast formed part of the Scotney Castle Estate and was oak framed, with part sandstone walling but mainly weatherboard clad, with an earth floor and a Kent peg tile roof. The original oast consisted of two 9ft square kilns with sandstone walls to a height of 4ft, then timber studding covered on the inside with wattle and daub, and finally weatherboarding on the outside up to the roof level.

In the 19th century the stowage was extended to an overall size of 30ft by 20ft and a 14ft diameter brick kiln added at the rear. In the early part of this century a second 14ft. roundel was built alongside the first. All four kilns were heated by coal (anthracite) fires. Charcoal produced on the farm and stored in a small lean-to wooden shed at the end of the oast was used to start the fires. The hops were dried on traditional floors, having timber slats covered with horse hair mats.

A wooden hand operated hop press, manufactured by G. Pierson, of Hurst Green, (see page 137), had been used until 1976 when hop production ceased on the farm. This unique press being over 100 years old was offered for sale in 1981. Robin sent his daughter to the auction with an "open cheque" to acquire the press for posterity, which he successfully achieved! The press was installed in the oast at the 'Museum of Kent Life', Cobtree, in 1984 and has been used every year since, during the annual Hop Festival in September pressing the hops grown in the museum gardens.

Having last been used to dry hops in 1976, the oast was sold in 1982 and converted into a domestic dwelling.

"HALES PLACE" TENTERDEN

The main feature of this oast is its very tall single cowl, previously with two vanes, constructed well above the roof ridge.

Unfortunately however during the past 10 years the cowl has lost the upper vane.

It was probably in the 18th century, when Sir Edward Hales rebuilt Hales Place, that the oast was constructed; the upper part of the wall having been built in Flemish bond. The 16th century English bond red brick wall adjoining the entrance gateway constitutes the lower part of the oast.

SWARLING MANOR OAST, PETHAM: BUILT 1790

This is a very large three-storey brick-built oast having six 16 feet square kilns positioned along the rear wall with two-storey brick-built extensions at both ends.

The oast is a listed building and has a number of unusual features. Each floor of the oast has a large unobstructed area of 100 feet by 24 feet. The green hops are raised to the top floor by a wooden hoist built within the roof space and then when required they are lowered to the kiln floor via a chute which is, in fact, a hinged door that flaps down ' supported by chains on both sides.

When the coal fires were removed in the early part of this century coke burning "Shew's patent economic pure-air heaters" (see page 112) were installed and an engine house was constructed to the rear of the oast. A "Blackstone" diesel engine coupled to a system of pulleys and flat belts was used to drive the fan installed at the base of each kiln.

September 1965 was the last time the oast was used for drying hops.

In 1983, following a heavy storm, four of the remaining five kiln roofs and part of the stowage roof collapsed, causing extensive damage.

In 1987 the stowage roof was restored and the oast converted into eight domestic dwellings.

ROCK FARM, NETTLESTEAD

This oast was constructed in two phases, with the second phase being an addition rather than a modification to the original structure. The original oast was built in 1830 by J. Hards of East Farleigh and had three 20ft by 18ft and one 10ft by 18ft brick kilns along the rear of the building and a timber studded front elevation covered with weatherboards. The two slate covered low pitch roofs had and still have hipped ends and a central gutter.

Centrally above each kiln along the roof apex small conical vents were constructed. These originally had wooden cowls but now are capped with metal covers. The low pitch and small roof opening would have restricted the natural escape of moist air and therefore also constrained the maximum depth of hops able to be dried at any one time. So at a later date a fifth kiln (measuring 22ft by 20ft) was constructed at the South West end which had a steep pitched slate covered roof. Initially a wooden cowl was fitted that allowed the reek to escape, but when a fan was later installed to further increase the air velocity through the hops, a louvred 'Beehive' type ventilator replaced the cowl.

When the working oast was first visited in August 1979 it offered a unique display of three different drying methods all under the same roof. One of the original kilns still had the hopper type (see page 108) firing although initially all the original four kilns would have been equipped with this type of drying. Another kiln had been equipped in 1946 with a hop drying plant installed by the Maidstone firm of Drake and Fletcher Ltd., (see page 114). Additions to the basic design consisted of an electric fan built into the outside wall at ground level to draw air into the kiln and also an automatic mechanical stoker (see page 115). When originally built the fifth kiln had the same Drake and Fletcher method installed. In 1963 it was converted to oil firing.All of the kiln drying floors were the traditional wooden slats covered with a horse hair mat. The hops were initially pressed into pockets using a wooden hand operated press built by B. Garrett, Maidstone. In the latter years a Drake and Fletcher metal press with a mechanical drive was used.

Since 1992 Robert Corfe has managed the farm, having taken over from his father. His family have been growing 18-22 acres of hops on the farm since 1965 and they have achieved many successes in the various hop drying competitions. Unfortunately due to market forces Robert Corfe has reluctantly decided that the 1994 season would be his last and the drying and pressing equipment have been sold.

The oast was sold in 1997 with planning permission for conversion to five residential units.

YONSEA FARM OAST, ASHFORD

This 160 year old Listed Grade II oast could be a possible casualty when the proposed rail link between London and the Channel Tunnel terminal is finally constructed. The oast consists of a 42 feet by 20 feet stowage with two 18 feet diameter kilns. The stowage walls are built of red bricks laid in Flemish bond with two Yorkshire sliding sash windows at the end on the upper floor. Its roof has a hipped end and is covered with slates with lead flashings along hips and ridge.

Two different bonds were used in the construction of the kilns. The first 10 feet consists of header only courses but the remaining 6 feet (up to the eaves) is in Flemish bond. Slates are used to cover the conical kiln roofs. The oast was last used to dry hops in the late 1940's and it is currently used as a garage and store for farm equipment.

SHEERLAND OAST, PLUCKLEY

A large ragstone oast which was built in 1838 for Sir Edward Cholmely Dering at Pluckley.

It consists of six 20 feet diameter roundels, attached to a 100 feet by 35 feet stowage. The stowage slate roof appears small, probably due to the span being divided into two, forming a secret gutter running the full length. Eight timber posts, equally spaced along the length of the stowage floor, support this gutter and the roof members, leaving the large cooling floor almost unobstructed.

The stowage has semi-circular headed iron window frames with brick reveals and ragstone elliptical relieving arches as well as elliptical ragstone arches to the six door openings.

Originally each kiln had a 3 feet 6 inches passage through the centre, with two coal fires on either side.

The cowl vanes have fixed to them large iron figures of the Dering crest - depicting a black staple shire horse.

PATRIXBOURNE OAST

The Marquess of Conyngham built this magnificent oast in the village of Patrixbourne in 1869.

The rectangular brick built stowage measures 75 feet long by 23 feet wide and has three 18 feet diameter brick roundels along the rear wall.

At the gable ends of the stowage, purlins and wall brackets support the unusual barge boards, which in turn support spear shaped finials. A date stone bearing his Lordship's crest is set into the brickwork.

Four gable end dormer windows and a lucarne (housing a pulley) are constructed into the roof at the front. All these dormers have upswept roofs, barge boards and spear shaped finials.

This oast was converted into two domestic dwellings in 1988.

GATEHOUSE FARM, MARDEN

This oast was originally built by Sibern Day Esq., in 1876 with two 20ft diameter kilns. The third roundel was added a few years later, as the hop acreage increased throughout the County. In 1912 a record 384 pockets of hops were produced, leading to the construction during 1914 of an additional 20ft square kiln at the other end of the stowage from the roundels.

At the front of the oast internal and external stairs gave access to the upper floor and at the rear an "endless" rope hoist was used to raise four pokes of green hops per lift. Hops were last dried over coal fires on the traditional slatted floor with horse hair mats in 1966. A hand operated metal press made by W. Weeks was used to fill the pockets. In later years the building has also been used to dry oats for the farms four shire horses, and also, as a sheep's wool weighing and grading centre. More recently the planning authority gave permission for the oast to be converted into a domestic dwelling.

This photograph of 'Gatehouse Farm' oast, was taken shortly after its construction in 1876. On the right Sibern Day is standing with his son and dog. Behind them pokes of green hops have been laid on sheep hurdles. Centrally a poke wagon is being unloaded. On the left a wagon stacked with pockets of dried hops ready to go off to market.

NASHENDEN OAST

The Wardens and Commonalty of Rochester Bridge, owners of Nashenden Farm, in the parish of Saint Margaret (now Borstal) Rochester, were requested in 1875, by the then tenant farmer Mr John Levy to build a two kiln oast.

Plans for a very substantial building dated 25th January 1876 where drawn up by Thomas Callund a surveyor. They were sent out to four builders inviting them to tender for the contract. Mr Thomas Nye of Star Hill, Rochester, returned the lowest tender of £770 and was awarded the contract and built the oast in 1877.

The oast was to comprise of two 20ft square kilns and a 50ft by 30ft stowage. A fire proof construction was specified for the first floor of the stowage. This consisted of 7" deep rolled iron girders, supporting 5"deep rolled iron joists, with a concrete infill, supported by wooden bearers. The upper surface was finished with a 2" thick cement screed.

Two rows of four cast iron 4" diameter stanchions, together with a 13½" thick perimeter brick wall, supported the weight of the heavy floor construction. The total weight was transferred to a load bearing subsoil by stepped brick footings and concrete pads or strip foundations.

On the farm the hops were always picked by hand into 5 bushel baskets. A daily lorry was sent out to the local villages of Wouldham and Borstal to collect the hop pickers, as there were no huts on the farm to accommodate pickers down from London.

A unique cast iron hop press was installed, manufactured by W. Weeks & Son, Maidstone, (see page 139). The press was never power operated and was last used to compress hops in 1981.

The 'Auger' family, grew hops on the farm for 66 years, from 1915 firstly with Frederick then his son Robert. Girl Guides regularly camped on this farm, in the beautiful peaceful Nashenden valley, during the late 1920's and early 1930's, sleeping and eating in the oast during bad weather.

After the second world war, up until the construction of the M2 motorway in 1968, cubs and scouts from the London Tooting troop, also camped on the site. Robert's wife Emma was awarded the Boy Scouts "Thanks" badge in the 1960's for her work with the young lads.

In 1984 the Rochester Bridge Trust sold the oast to Messrs W. H. Simmonds & Son, Builders of Wrotham, Kent, who converted it into two spacious four bedroomed houses in 1988.

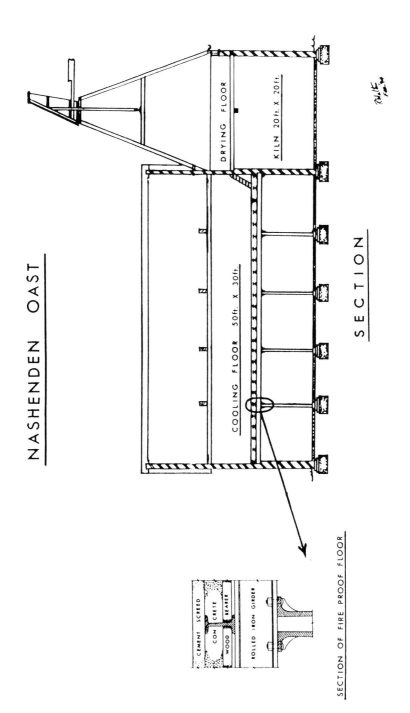

NASHENDEN OAST

DRYING FLOOR

KILN 20ft. x 20ft.

COOLING FLOOR 50ft. x 30ft.

SECTION

CEMENT SCREED

CONCRETE

WOOD BEARER

ROLLED IRON GIRDER

SECTION OF FIRE PROOF FLOOR

SPRING GROVE FARM, WYE

In 1898 a fire destroyed two of the three 18 feet diameter kilns which formed the oast on this site. The two roundels destroyed were replaced the following year with 20 feet square kilns.

It became uneconomical in 1980 to continue to use all three kilns to dry the 36 acres of hops remaining on the farm. The solution was to consolidate on one of the square kilns by removing the Hetherington roller hair floor in the left hand kiln and then converting it into a three tier drying system, the only one known to exist at that time in Kent.

Green bags (pokes) were mechanically elevated to the top floor at eaves level. There the hops were spread out over a metal louvred floor and warm moist air from the two drying floors below passed through the green hops slightly reducing their moisture content.

The next stage was to open the louvres allowing the hops to fall onto another louvred floor below, where further drying took place from the less humid air passing through them.

Finally the second louvred floor was opened to allow the hops to fall into six bins at the first floor level, where the 8 hour drying period was completed. After which wooden shutters were raised to allow the bins to be wheeled out on to the stowage floor and subsequently the hops, using a hydraulically operated press, were formed into bales.

This three tier drying system was very economical with only about half the amount of diesel fuel required compared to the previous three single floor kilns.

The farmer, Mr Charles Amos, decided in 1986 that it would be his last year of hop farming.

On Saturday, 6th September 1986 the farm held an Open Day to raise funds to repair the Wye Village Hall. The weather was fine and a large number of people attended. Shortly after this event the oast was converted into holiday flats, managed by the farmer himself.

The converted oast was officially opened by The Baroness Trumpington, Parliamentary Secretary in the Ministry of Agriculture, on the 5th August 1987.

In August 1994 the oast was sold as six private apartments each named after varieties of hops; Pride of Kent, Brambling, Early Bird, Eastwell, Northdown and Brewers Gold. Thus bringing this era to an end. The Amos family had grown hops on this farm for more than four generations.

Spring Grove Farm, Wye.

Boughton, Faversham.

71

BOUGHTON FIELD OAST, MACKNADE FARM, FAVERSHAM

This four square kiln oast, built in 1896, that stands alongside the A2 road, was converted in 1985 into eight dwelling houses.

In 1919 Leslie Christopher Dunk, as a 19 year old lad, started work for Sir Thomas Neames at Boughton Field oast. When interviewed in 1984 at the age of 84, he could still recall what it was like, having just been released from the army following the 'Great War'.

Chris, as he was commonly known, came to the farm to work with his father (David) whom he succeeded as chief drier in 1936, until 1967 when he retired.

Chris recalled that his first job at the beginning of the hop-picking season was to remove all the old swallows nests, clean the floors and, most important, to fill a pocket full of straw to act as a comfortable bed ready for the long night vigils.

A cast-iron kettle was placed on a kiln fire in readiness for the midnight feast, consisting of tea and 'jacket' potatoes cooked in the ashes of the kiln fire, sliced in half and coated with butter or beef dripping.

Originally all four kilns were heated, using coal fires, but in the early 1900's a 'Sirocco' burner (page 118) was installed at the rear of the two left-hand kilns and warm air was forced into the kiln from below the traditional wooden slatted floor covered with a horse hair mat.

By contrast the two right-hand kilns were converted initially to coke-burning 'Cockle' stoves. Later in the 1940's oil burners made by Drake & Fletcher of Maidstone were installed. The drying floors of these two kilns were of the roller hair cloth type.

Inspection of the drying hops was also carried out differently in the two pairs of kilns. A roller platform was used in the left-hand pair and rotating platforms suspended from a roof support beam in the other pair.

Two wooden presses made by H. & F. Tetts of Faversham were used to fill the pockets.

BULL LANE, BOUGHTON, FAVERSHAM

"UNLOADING GREEN BAGS", BOUGHTON, KENT.

This scene is one of a series of postcards printed by Young and Cooper of Maidstone in 1905 covering the Boughton area.

It depicts pokes being hoisted up into the roof space of the oast sited at Vine Farm, Boughton and currently owned by Allan Bones Ltd.

This oast was last used for drying hops at the end of the second World War. It is now used as a fruit store, but the three kiln roofs have been removed, as have most of the louvred windows, and the openings 'bricked in'.

Hops are still grown in the adjoining gardens. They are picked by machine and dried, using oil-fired burners, in a more recent oast on the opposite side of the road from the original oast shown in the postcard.

CLOCKHOUSE FARM, HUNTON

This particular oast built in 1928 was the last one to be constructed in Kent with the Victorian style of revolving cowl on the kiln. After this date the more modern louvred ventilators, running along the ridge, were used.

The original oast, with a single 16 feet diameter roundel, was destroyed by fire early in 1928 and the remains can be seen in the foreground of the accompanying photograph.

The three-storey oast comprises a single 18 feet square kiln and a 20 feet by 18 feet stowage and has walls of 9 inch thick brick laid Flemish bond with both the roofs covered with clay tiles. The inside of the kiln roof is lined with asbestos cement sheets in lieu of lath and plaster for added fire protection.

An electric hoist hauls the hop pokes up to the top floor of the stowage and the hops are then spread out in the kiln, over a horse hair mat laid on the wooden slatted floor.

Originally the kiln was coal-fired but in 1963 a coffin-shaped oil-fired burner was installed by Messrs. W. Weeks and Son, Maidstone, with a cut-out alarm bell which has subsequently been uprated with a heat sensor device. A "puller" fan fitted in the roof of the kiln expels the reek from the hops.

The dried hops are pocketted using a "Weeks" wooden press that was motorised in 1957.

During the war years (1939-45) the Day family set up two 'Spitfire' bins, where the pickers gave their services free and the proceeds contributed towards the war effort.

Mr Peter Day grew 10 acres of hops on this farm, as did his father before him. A band of 'strangers' (Londoners) came down each season to hand pick the crop. This is one of only two farms in the county that still picked by hand until 1985. The same families had been involved in this 'holiday' activity for the past century.

In this modern scene taken in the 1982 season, Peter Day is seen in the role of measurer (centre). The jovial, bespectacled man perched on the left hand end of the bin is George, from Billingsgate, London, who at 84 years of age still enjoyed his annual visit to the hop gardens which he had done for as long as he could remember.

SAYNDEN FARM, STAPLEHURST

In 1949 two Merchant Bankers redeveloped the remains of a two kiln oast at Saynden Farm, Staplehurst.

New 9 inch thick stowage walls were built in red sand faced bricks to blend in with the original paler red brickwork of the kilns and incorporated steel window frames with pivot hung sashes. Large asbestos cement sheets cover the roof and these are supported by steel purlins and roof trusses.

A new 20 feet square kiln was constructed and this together with the original 16 feet diameter and 14 feet square kilns were covered with asbestos slates, those covering the two square kilns being laid unusually in a diamond fashion.

The weight of the large 45 feet by 25 feet unobstructed stowage floor is carried by a single 22½ inch square brick pier which in turn supports the 12 inch deep steel beams that carry the wooden floor joists and boarding. At the rear of the oast is a 10 feet wide timber gantry on which the pokes of green hops were placed before being tipped out on to the kiln floor.

Messrs Drake and Fletcher, (Maidstone) installed 'modern' coke burning furnaces with an automatic hopper feed unit in all three kilns. Each kiln had a traditional slatted floor covered with a horse hair mat. When I visited the oast (in 1994) the hop press had been removed, but I was informed by the owner that a wooden framed press with a mechanical drive attachment had been used. It was the common practice in most oasts to stencil on the wall alongside the press the number of pockets pressed each year. This oast recorded 153 pockets in 1949 its first year in operation.

The oast was last used to dry hops in 1967, a revived working life of only 18 years.

PORTABLE OAST

A fire destroyed the oast at 'Great Pedding Farm', Wingham, near Canterbury, during the hop drying operations in 1992; but this did not stop the farms long history of hop production.

For the 1993 season a portable hop drying system was introduced, which is still in use today. It consists of 12ft square plywood bins with a wire mesh bottom, into which the green hops are loaded. Each full bin is stacked on an empty bin, into which hot air is blown from gas burners. A sensor in the lower bin connected to a control box on the burner, monitors and adjusts the temperature.

When the hops are dry the full bin is raised and the hops tipped into a stowage bin to cool and obtain a state of equilibrium of 10% moisture content.

A conveyor belt carries the hops from the storage bin to the two automatic hop presses. Hop pockets are suspended in an empty steel "water" tank below ground level and when full are sewn up and then lifted from the tank by a mechanical hoist.

At the end of the hop drying season the elevator, bins and gas burners are stored away in a corner of the building. Only the hop presses and the underground chamber remain in place, leaving the majority of the floor area available for other uses.

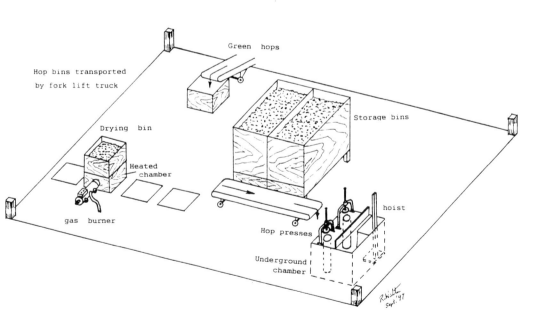

Chapter 3

Extremes of shape and size

THE OLD OAST, HAWKHURST

Kilns of an unusual shape

A pair of octagonal kilns, with 15 inch thick sandstone ashlar walls with rooms 15 feet across the flats of the octagon.

These kilns were built on to the end of a rectangular stowage which has a sandstone ground floor wall and timber weatherboarding to the upper storey.

Purpose made angled clay tiles were used to cover the hips of the kiln but these special handed tiles are no longer obtainable. Recently the right-hand kiln had to be retiled and it was necessary to use the normal tapered clay tiles, mitred and bedded in mortar at the hip intersections.

OUTRIDGE FARM, BRASTED CHART

The National Trust own a unique ragstone oast, built in the early 19th century, at Outridge Farm, Brasted Chart.

As this oast was built on sloping ground, the added height of wall, at the lower end, has enabled arches to be built into the ragstone providing cattle shelters under the floor level of both kiln and stowage.

The three 18 feet square ragstone kilns have tiled roofs and each has an octagonal shaped cowl. The cowls are enclosed on six sides with vertical boarding with an octagonal cap and knob finial, with the vane terminating in a crosslet.

Timber weatherboarding is used to cover the stud timber framing of the upper storey of the stowage above the ragstone level. The roof is supported on 'King Post Roof Trusses'. In this oast the 'King Post' is an iron rod though normally it is a timber member.

LARGEST CLUSTER OF OASTS

The largest group of oasts in the Country are reputed to have had a total of over 30 cowls, and were sited at the northern end of 'Herts Farm' on the Peale Estate, Loose, near Maidstone. When the Peale Estate was sold on Thursday, 13th June 1912 most of the oasts and allied buildings that once dominated the village had been demolished.

It was most probably following the 'Mass Demonstration of the Hop Farming Industry' in Trafalgar Square, London, on Saturday, 16th May 1908, when farmers demanded a 40/- per cwt tariff on imported hops, which the government of the day would not impose, that a large number of the oasts on the estate were demolished and the remainder converted to other uses. Also the hop gardens became grazing land, fruit orchards and building plots.

Only two of the original oasts remain:

Sale Lot No 25.
An oast that had been converted into five substantial cottages called 'Oast Cottages', these have since been renamed 'Fairview Cottages'. At the rear are several large stone built cellars. These are excavations in the bank and formerly used for storing coal to fire the kilns.

Sale Lot No 26.
From the Bill of Sale—"A most substantial building, 70ft by 27ft 6in, constructed of stone, laid in courses and partly timbered, with an excellent King-Post slated roof, formerly used as an oast, having four spacious floors, supported by massive timbers and a basement. Very suitable for a Warehouse or Factory".

This property is now a Listed Grade II building in the grounds of the 'Old Wool House'. Colonel J.C.B.Statham purchased the 'Old Wool House' and Oast on his retirement in 1921 transforming the barn-like building (oast) into a private museum for his trophies. At his death in 1932 the 'Wool House and Cottage' were left to the National Trust.

Unfortunately, the oast building is now just a shell of the original structure covered by a mono pitch corrugated iron roof.

BELTRING HOP FARM

Largest Group of Victorian Oasts
The largest existing complex of oasts can be seen at the former Whitbread Hop Farm, Beltring, where four oasts, each with five brick roundels complete with cowls, provide a picturesque scene. Many passers-by stop on the roadside to admire their splendour and count the cowls.

Siegfried Sassoon wrote in his book, 'Memoirs of a Fox-Hunting Man', published in 1928 by Faber and Faber Ltd, London:

"It was unusual to find more than two hop-kilns on a farm; but there was one which had twenty and its company of white cowls was clearly visible from our house on the hill. I would count them over and over again, I felt that almost anything might happen in a world which could show me twenty hop-kilns neatly arranged in one field."

This magnificent group of late Victorian oasts was the work of a local farmer, Mr E.A.White, who owned the land prior to Whitbread purchasing it in 1920.

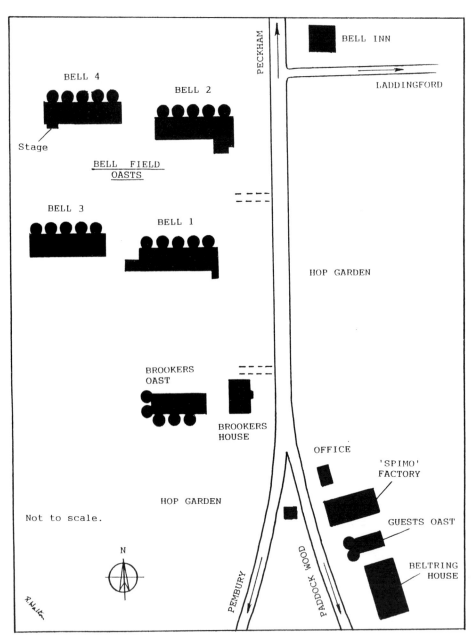

Beltring 1918

Later Developments

During the past 150 years there have been a total of fourteen different buildings used for hop drying on the farm, and it is interesting to note that no two oasts have been identical in construction, drying techniques and pressing equipment.

Over the years Whitbread have invested in new technology to keep competitive with foreign imports. This culminated in the most advanced handling, drying and packaging techniques incorporated in the new oast, which was built behind 'Lily Hoo' house in 1985 and eventually sold in 1991.

REF. OAST	STOWAGE			KILN			LAST USED FOR DRYING
	SIZE	STOREYS	PRESS	NUMBER	SIZE	FIRE	
1 Bell 1	26' x 100'	2	Metal	5	20' dia.	Coke	1946
2 Bell 2	26' x 100'	2	Metal	5	20' dia.	Oil	1984
3 Bell 3	29' x 110'	3	Metal	5	20' dia.	Oil	1984
4 Bell 4	29' x 110'	3	Metal	5	20' dia.	Coke	1980
5 Bell 5	74' x 107'	2	Metal	5	20' sq.	Coke	1973
6 Guests	19' x 40'	2	Wooden	2	18' dia.	Coal	1958
7 Brookers	20' x 63'	2	Wooden	3	16' dia.	Coal	1948
				2	20' dia.		
8 Lily Hoo	45' x 50' approx.	2	Wooden	2	16' dia.	Coal	1942
				1	18' dia.		
				1	20' dia.		
9 Stilstead	28' x 86'	3	Metal	4	20' dia.	Coke	1983
10 Barnes St.	30' x 80'	3	Wooden	3	20' dia.	Coal	1966
				1	18' sq.		
11 Long Oast	Brick and timber structure behind 'Beltring House', last used early 1900's						
12 Fan Oast	Brick structure next to 'Brookers' oast, demolished in 1923.						
13 Experimental Drying	Timber structure behind 'Lily Hoo' oast, destroyed by fire in 1947						
14 New Drying Shed	Steel framed building behind 'Lily Hoo' House, constructed in 1985						

Guests Oast, Beltring

This oast is one of the oldest still standing on the farm and named by E.A.White after his chief drier- Henry Guest.

It was mainly used for experimental purposes. Triangular shaped wooden bins with perforated bases were placed on the slatted drying floor, radiating out from the centre of the kiln. Each bin contained a known quantity of hops from a specific variety, the aim of which was to compare their characteristics under similarly controlled conditions.

A 'Weeks' wooden hand-operated press was used to fill the hop pockets. The stowage comprises of brick ground floor walls, with the upper storey timber framed using re-claimed oak and covered externally with tarred softwood feather-edged weather boards. Initially wooden, horizontal pivoted, louvred windows were installed for lighting and ventilation. The roof trusses are also of re-claimed oak and originally covered with hand made clay peg tiles.

The two 18 feet diameter roundel roofs were stripped and reconstructed in 1980 and at the same time sky lights were inserted. Also external doorways were constructed in the ground floor brick walls and where once coal burning brick hearths stood, one of the Whitbread Company's shire horses was housed until 1991 when the oast was sold.

Long Oast, Beltring

This building is located behind Beltring House and comprises of brick ground floor walls and a timber-framed, weatherboarded upper structure.

Two small dark coloured bricks, set into the wall above an enlarged front entrance doorway are engraved 'TWC 1823'. The Yalding Tithe Map of 1840 shows the Cheesman family in residence during this time.

Thomas and William, who leased the farm from the Drapers' Company probably built the oast in 1823. The building layout is typical of a barn converted to an oast in the 19th century, in that it contained both the stowage and kiln under one roof. The size of the oast is deceptive, since the roof has a valley running front to back. Below this valley, both at ground and upper floor level, there was a 4 feet wide passage, Either side of this passage, on the first floor, were large rectangular conventional open timber slatted drying floors.

The 'kilns' would have been loaded from the passageway, and the dried hops discharged down a chute at the end of the passage, on to the cooling floor at ground level. The dried hops were scuppetted into a pocket, which was suspended into a pit sunk into the ground. In 1981 the ground floor was modified to provide stalls to house the Whitbread Company's shire horses and in 1991 the oast was sold. It has now been converted into a domestic dwelling.

LARGE OASTS

Three large identical eight kiln oasts were built on 'Court Lodge Farm', East Farleigh in the 1870's.

These oasts consist of eight square kilns, built in 9 inch thick brickwork, laid Flemish bond. The 18 feet square kilns are arranged along three sides of a 75 feet by 50 feet stowage. The stowage is constructed of 13½ inches thick brick walls laid Flemish bond, with weatherboarding above the upper floor window at the gable end. Doors are the heavy timber-frame, ledged, braced and batten type and windows are rectangular wooden casements.

Access to the upper storey is by an internal staircase. The 75 feet upper floor span is divided into two by a row of cast iron supports at 10 feet centres which carry the weight of the valley and roof members. Each roof is supported by Queen Post Roof Trusses spaced at approximately 10 feet centres across the width of the stowage. British Columbia Pine 38 feet long, 12 inches deep and 6 inches thick is used for the tie member of the trusses.

Both kiln and stowage roofs are covered with slates and lead is used to weatherproof at intersections of hips, valleys and abutments to walls.

In 1932 the three oasts, accompanying 225 acres of celebrated hop land and 104 hopper huts were all put up for sale at the Royal Star Hotel, Maidstone, by the owners, Kent Hop, Fruit and Stock Farms Ltd. The three lots were purchased by different local farmers and the following specifications are all taken from the original bill of sale.

Churchfield Oast

'This has a steel gantry for storage of green hops and fitted with "Hetherton's" roller hairs and electric fans. An engine house and dynamo shed of brick, timber and felt construction houses a 12 h.p. "Lister" petrol engine for power and lighting. Sixty nine brick and tiled or timber and tiled hopper huts and a cookhouse. Also a range of timber and corrugated iron privies with flushing systems.'

This oast was last used to dry hops in the 1977 season and was still in very good structural condition, but unfortunately only sixteen of the original brick hopper huts remain. The oast was sold in 1982 and has been converted into four domestic dwellings.

Coombe Bank Oast

This lot comprised:- 'The oast, together with approximately 17 acres of hop land and 35 brick and tiled or timber and tiled hopper huts. Also 2 brick and tiled cookhouses each with 30 fireplaces and corrugated iron shelters, and a range of timber and corrugated iron privies with flushing system.'

86

On the 9th April 1987 during a violent thunderstorm the oast was struck by lightning and the ensuing fire completely gutted the building. The building at the time of the fire was being used for the storage of apples. Hops were last dried in the oast in the early part of this century. Today all that can be seen on the site is a single 18 feet square kiln and accompanying stowage built in September 1987 on the rear walls of the original oast.

Court Lodge Farm Oast

The third oast no longer exists. On the 6th May 1955 a fire started in the upper structure of the oast and it quickly spread to the adjoining buildings and for three hours scores of firemen, using eight pumps called from different parts of the county fought the huge blaze.

The two remaining oasts can best be viewed from the public footpath behind the East Farleigh railway station car park. Also from this spot the local church can be seen up on the hill. In the adjoining graveyard, next to the war memorial, a wooden cross marks the place where hop-pickers were interred having died of cholera, whilst in the employ of a Mr Ellis - a large hop grower. The vicar of East Farleigh at the time took immediate steps to check this outbreak. He was the Rev. W. Wilberforce, the son of the statesman. The 4ft 6in high cross has engraved upon it the words:

"In Memory of Forty-three strangers, who died of cholera,
September 1849. R.I.P."

Church Field Oast

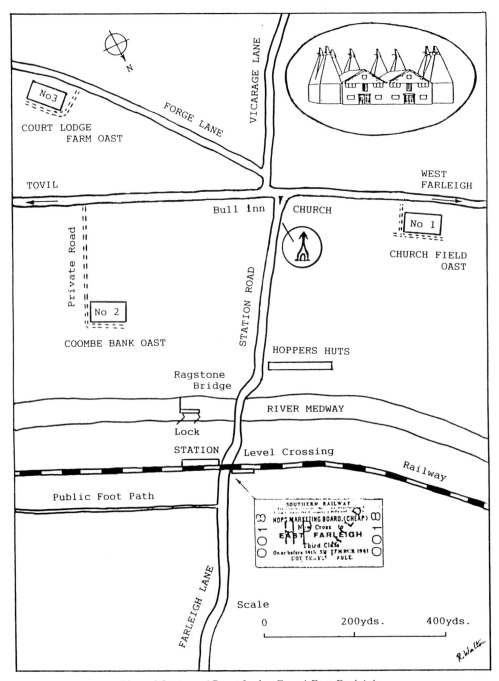

Plan of Oasts on 'Court Lodge Farm' East Farleigh

Crowhurst Oast, East Peckham

This large eight-kiln oast was built in 1884 in the village known as "Peckham Bush". It consists of two 14 feet diameter roundels and six 18 feet diameter, with a three-storey stowage measuring 80 feet by 24 feet. All the kilns were originally brick throughout, i.e. both walls and roofs. Tiles were later used as a roof covering after it was found that water had penetrated the structure. In the front right-hand kiln in 1942 some hops caught fire and damaged the roof. This is the only one of the eight kilns that now has a timber roof.

Another unusual aspect of this oast is the 'Yorkshire Light' windows with wooden horizontal sliding sashes.

The farm has been managed by three generations of the Dolding family. James Dolding was the first, from 1902 until 1933, when the responsibilities passed to his son George, who remained manager until 1949. Charles started work on the farm in 1926 at the age of 14 and later took over from his father until the farm was eventually sold in 1969.

Aerial View of Crowhurst Oast

A staff of four - one head drier, one coal stoker and two pressers - manned the oast. Coal fires were used to dry the hops with two 'Weeks' iron frame presses to fill the pockets and hurricane lamps for lighting. Electricity was installed in 1938. On a nearby site a second oast was built in 1906. This was a two-storey building with four 18 feet diameter roundels. In 1965 it was burnt down and destroyed. A new brick oast now stands on the site.

A book entitled, 'Saint Francis of the Hop-Fields' was published in 1933 and written by the Rev. Miles Sargent, who tells of his work amongst the London hop-pickers living in the 125 huts on this farm. He started his Mission on the farm in 1926 and was helped by a group of medical under-graduates from Oxford.

In the accompanying photograph the corrugated iron building behind the van was the canteen, which had a theatrical stage and the wooden building situated on the right was the church.

The emblem of the group in the photograph was a red and white handkerchief, which each member normally wore around their necks. This Mission finally closed in the 1960's. The oast was last used for drying hops in 1973 and from then until 1981 it was used as an apple store. It has now been converted into four domestic dwellings.

The Crowhurst Farm Mission

SMALL OASTS

This small oast is at Bethersden and was built in the 19th century. It consists of a single 12 feet diameter brick kiln with 4½ inch thick walls laid stretcher bond. Headers appear where the piers and corbels are built into the wall to support the timber plate on which the drying floor joists rest.

This single kiln is built at the end of a weatherboard timber-framed stowage measuring 24 feet long by 15 feet wide. At the opposite end, a pair of ledged, braced and battened doors are hung, to enable the hop pockets to be carried out for loading and despatch by horse-drawn cart or lorry.

A 'collar' type roof covers the stowage and a conical timber roof the kiln, both covered with hand-made clay tiles held in position by wooden pegs.

This was converted into a domestic dwelling in 1983.

Chequer Tree Oast

On the outskirts of Marden village, at Marden Beech and Wanshurst Green, can be seen small single kiln oasts, which have now been converted into domestic dwellings.

The kilns are approximately 11 feet in diameter, 13 feet from ground level to the eaves and constructed of 9 inch brickwork laid to a Flemish bond. The roofs are tiled and in both cases a new cowl has been fitted.

Wanshurst Green Oast

Chapter 4

Building Conversions

UTILISATION OF EXISTING ANCIENT BUILDINGS

Over a period of 350 years (1550-1900) during the expansion of hop farming in this Country, many hundreds of purpose built and converted buildings were used for drying, pressing and stowage of hops.

In the build up to the peak production, in the late 19th century, many very old buildings were utilised for hop stowage with often a kiln built adjoining it or one built within the original structure.

Saint Botolphs Church, Ruxley Manor, Sidcup, (O.S. map ref. TQ485702) is one example.

A 13th century Parish Church, built of flint, rubble and chalk blocks was desecrated in 1557. It was later used as a barn and an oast was built onto the East corner, consisting of a single 16ft brick roundel under a tiled roof, complete with a traditional cowl.

The church has been declared an Ancient Monument and English Heritage have donated money to its restoration. Both the church wall and roof were repaired in 1993 but the roundel, which until the 1987 hurricane still had a cowl, is now in ruins, with only part of the brick wall remaining.

Another very old building that was used as an oast is the 14th century 'Grand Hall' at Charing Place, where a square kiln has been built within the hall. By standing against the Lych Gate of Charing Church, the apex of the kiln roof can be seen when looking in a southerly direction.

'Hawkenbury Oast', near Staplehurst, although it appears from the roadside to have been built within the last 150 years, has a central stowage structure that is probably the same age as 'Hawkenbury House' alongside, which is actually a 16th century building. Evidence to suggest this is the 'Crown Post' supporting the stowage roof. Also there are signs of soot from possibly an open fire and plaster from an upper room partition.

A planning application was successfully submitted for conversion of the oast in 1995. This building is unique and was once a 16th century domestic dwelling, converted to an oast and now back to a modern domestic dwelling.

BIDDENDEN: BUILT 1540

The original building was a rectangular timber-framed barn, 40 feet long, 20 feet wide and 12 feet high at the eaves, with a hipped hand-made clay peg tiled roof.

The timber stud walls were covered, on the inside with wattle and daub and on the outside with timber boarding.

In the late 18th century the building was extended by 6 feet in length (the old hip rafters still remain) with a cowl fixed into this extended roof. An upper floor erected and windows introduced at both levels. A ground floor wall was built in 9 inch brickwork (Flemish bond) to form an 18 feet square internal kiln.

In 1980 this timber-framed building was clad in brickwork and converted into a domestic dwelling. The small cowl can still be seen today projecting from the extended roof.

 **CATT'S PLACE, Nr MILE OAK**

This timber-framed building was originally constructed in the 16th century as a barn and later converted to an oast.

During the 18th century considerable internal alterations were made resulting in an unusually large rectangular kiln 27 feet long by 18 feet wide with brickwork up to the eaves level. This has been divided into two, each with a cowl built on to the ridge.

The ground floor consists of a central arched passageway running throughout the length of the brick lined area. Four brick furnaces were built on each side.

A slatted upper floor has been constructed above the furnaces and the roof space lined with timber laths then plastered. A door leads out onto the cooling floor.

A hop picking token dated 1774, found in this area, further supports that hops have been grown on this farm for at least 200 years.

BARNHILL FARM, HUNTON

Dated: 1623 on Beam

This is an example of a typical Jacobean gable, with original very high quality carvings on the barge boards and carved animals on the beam.

This very old Jacobean building was converted into a stowage in the 19th century with two roundels (of 9 inch thick brickwork laid in Flemish bond) built on to one side. On the opposite side, a farm cottage was built with the carved barge boards of the porch fixed above the doorway.

Both the kilns and stowage fell into disrepair, but in 1986 the stowage was converted into a domestic dwelling and part of an oast barge board, with its original 17th century carvings was used internally as a dado rail in the lounge.

The two roundels however were demolished and the area now forms two circular patios.

CHIDDINGSTONE

This superbly converted oast has retained, and exposed, the oak framed wall and Queen strut roof members. It is a listed building within a conservation area.

The original building was an oak framed weather-boarded barn, with hand-made clay roof tiles and was probably constructed between 1750-75.

In the early 19th century the barn was converted into the oast. A square kiln actually constructed within the north end of the converted building and a cowl built into the roof apex are both typical of this period.

In the 1870's a brick extension was built onto the south end, together with four circular brick kilns. Within each roundel a timber roofed hopper fireplace was constructed, with four coal fires in each. A wooden 'Weeks' hop press was installed in the stowage area.

It was last used for drying hops in 1944 and converted into a domestic building in 1975-6

STILE BRIDGE, MARDEN

This oast was originally a small 29 feet by 18 feet farm outbuilding. Comprising of 9 inch thick brick ground floor walls, with a timber framed upper story covered with hanging tiles on the outside and internally with tongue and grooved boarding. The roof had jerked (barn) hip ends and a collar type construction with a single 8 inch square tie beam that gave only 5 feet 6 inches headroom. Its rafters were reclaimed oak halved together and pinned at the apex, with no ridgeboard, having a covering of Kent peg tiles bedded in lime mortar.

A brick wall divided the earthen ground floor into the kiln and stowage areas. Both having ledged doors to give direct access from the outside. Along the 18 feet east end wall two square hopper type coal fires were constructed. The 18 feet by 9 feet traditional slatted drying floor above had been raised 3 feet to give an overall height of 9 feet 6 inches, the normal height being 12 feet. The inside of the kiln roof was lathed and plastered, but not trimmed, to form an opening for the small conical vent which was capped with a square metal cowl. Access to the upper story was by an external flight of steps, to a squat 5 feet high by 3 feet 6 inches wide ledged door leading onto the pressing floor.

The hop press was an early wooden model from W. Weeks of Maidstone. The 11 inch by 3 inch side members continued down to the ground floor and a saddle (to support the pocket) was positioned in grooves and raised or lowered by steel rods connected by chains to the press. Stenciled on the wall alongside the press was the date 1877, which was possibly the year that the building was converted to an oast. Also the number 11 was visible, probably indicating the number of hop pockets filled that year.

This oast was converted once again, this time in 1988 into a domestic dwelling and a small white traditional wooden cowl installed.

CONVERSIONS FROM AN OAST TO OTHER USES

Present Use of Oasts

Of the hundreds of oasts built throughout Kent from the 16th to early 20th century, only 72 are still being used for their original function, i.e. hop drying.

Many of those still standing are used for storage of fruit and farm machinery, whilst others have been converted into domestic dwellings. Local authorities and preservation societies, in an effort to save the oasts, have applied to the Department of the Environment to place preservation orders on certain buildings, to prevent their alteration or demolition.

In recent years, a growing number of oasts have been sold for conversion to domestic dwellings. The local planning authority is conscious of the need to retain the original outline of the building and will not allow indiscriminate dormer window construction in the kiln roof. They also specify that a cowl must be maintained on the kiln.

For those wishing to take a holiday in Kent and stay in an Oast, there are many bed and breakfast or full accommodation privately owned oasts available. Addresses can be obtained from local Tourist Information Centres.

If you are contemplating a longer stay, the local newspapers often carry advertisements of oasts to-let.

Tourist Information Centres at:

Maidstone, Royal Tunbridge Wells, Sevenoaks, Tonbridge and Westerham. Cranbrook (seasonal only).

Accommodation offered in 1997.

Bed & Breakfast and Guest Houses at:

Bearsted, Biddenden, Chiddingstone, Cranbrook, Hadlow, Hever, Hunton, Lamberhurst, Langley, Paddock Wood, Sutton Valence, Tonbridge and Yalding.

Self Catering at:

Goudhurst, Hildenborough and Sutton Valence.

Meopham

In the 1820's an oast consisting of a single roundel was built on Millars Farm, Meopham. It had knapped flint walls under a tiled roof.

In 1903 Sir Philip Waterlow had the oast converted into a domestic dwelling, incorporating a total of thirteen windows, of which ten were dormers or partial dormers, built into the roundel roof.

This building still remains virtually unchanged from the original conversions and has been renamed 'Tower Folly' by the present owner.

Ye Olde Hop Oaste, Meopham

Apart from purely domestic conversions various other uses have been found for redundant oasts.

Tonbridge boasts a theatre housed in a two roundel oast opened on 20th April 1974 by Lady Rupert Nevill.

Whilst Rainham has an Oast House Theatre and also a Community Centre that is housed in an old oast.

Two towns museums have sections devoted to the history of hop picking. These are located at Tenterden and Cranbrook.

In Preston Street, Faversham, a 16th century building has been converted into a Heritage Centre and Museum which includes a hop display.

Oast Theatre, Tonbridge

Community Centre, Rainham

Oast House Theatre, Rainham

HOSTELS

Ellenscourt Oast

A small 19th century oast on the Lady Margaret Manor Estate at Doddington near Faversham has during the past 90 years been used in various guises as a haven for young and old people.

At the turn of this century, Dr Josiah Oldfield, a leading dietetic specialist, bought the almost derelict oast and converted it into a private hospital.

In 1942 it was purchased by the Stansfield Association, formed by two gentlemen Roy Stansfield Ashby and Jim Babbington who devoted their lives to helping the children from the slums of London to achieve a better way of life.

From 1947 to 1980 the Youth Hostel Association used the oast as an overnight stop for walkers and cyclists touring the North East of Kent. Ken Parfitt was the warden here for 25 years.

For a short period of time in the early 1980's it became the Charles Darwin Field Study Centre.

It is currently managed by Mortimer Homes Ltd, whose aim is to help people with learning difficulties.

POST CARD

Address only this side

DODDINGTON HOSTEL.

" An Oast House, originally used for drying of Hops, one of Kents specialities . . . then converted into a Private Hospital . . . later on, during the last War, used as a Boys' Home for lads from London's East End . . . now, as a Hostel, awaits to Welcome YOU into its unique atmosphere from all corners of the Earth . . . " Location 1½ miles S. of Doddington Village.
O.S. No. 172.   Barts ½ inch No. 10.   Nat. Grid 925552
KEN PARFITT, Warden.

Cira 1947

Front View 1996

Ellenscourt Oast

Capstone Farm, Rochester Upon Medway

Capstone oast was last used for drying hops in 1946. For the next 50 years it was used firstly as a store for farm machinery and cattle food and in latter years as a stable for horses.

In 1996 the Rochester City Council funded the Conversion of the two kiln oast into a 41 bed Hostel and leased it to the Youth Hostel Association (YHA).

Councillor Mrs Julie Shaw officially handed over the keys to Derek Hanson, YHA Chairman, on 12th July 1996. This event was witnessed by about 35 people, opening the doors to the youth of the world, giving them the opportunity to visit the historical City of Rochester.

Chapter 5

Internal Workings

COAL FIRES

Single Fire

The simplest type of open coal fire consists of a single, large rectangular, brick-built hearth with a sheet of corrugated iron suspended above to deflect the heat evenly throughout the kiln.

Some designs have the top closed in by brickwork or concrete and in these the brickwork of the upper part of the sides is 'honeycombed' i.e. leaving out bricks at intervals, so that the heat and fumes may escape.

Multiple Fire Layout

Across the middle of the lower storey of each kiln is an arched brick-built passage about 4 feet wide and 7 feet high at the apex of the roof. On each side of this passage are 2-4 brick-built hearths for the open fires, each approximately 1'-9" wide and 3'-0" long.

The fire bars are about 2 feet above the floor of the kiln and the fires are stoked through brick openings or cast iron doors from the passageway.

A small ventilation opening fitted with an adjustable wooden shutter is set into the passage wall adjacent to each hearth opening to control the cold air flow. The heat passing upwards through the kiln can therefore be controlled by varying the quantity of fuel used and by adjusting the passage ventilators.

The wooden door at the end of the passage, which gives outside access to the kiln, is often fitted with adjustable wooden louvres providing additional heat control.

Also on each side of the passage by the entrance there is a wooden door which gives access to the space between the passage and the kiln wall.

This type of layout is found in both round and square shaped kilns.

COAL FIRES

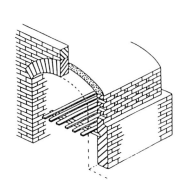

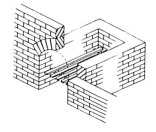

Open Coal Fire

Closed – in Coal Fire

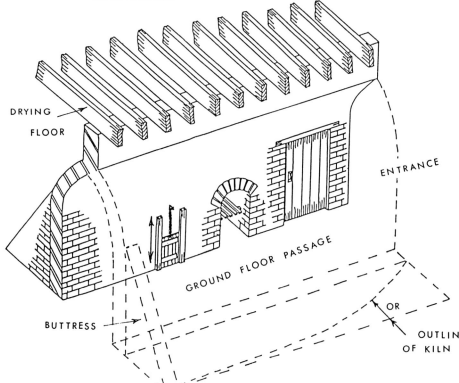

DRYING
FLOOR

ENTRANCE

GROUND FLOOR PASSAGE

OR

OUTLINE
OF KILN

BUTTRESS

R Walton

Multiple Fire Layout

Hopper Construction

In an attempt to obtain an even distribution of heat under the drying floor of square kilns, a group of four coal fires were constructed in the centre of the ground floor. The plenum chamber was formed in the shape of an inverted pyramid, constructed of a dwarf brick wall roofed with timber rafters, laths and plastered on the upper side, with clay roofing tiles bedded into the plaster to give a more durable surface. This type of structure within a kiln was known as the 'hopper' pattern. The fires were tended from the passage within the kiln formed by the 'hopper' type construction.

The 'hopper' principle was also used in round kilns, with three or four coal fires sited in the centre, surrounded by a dwarf wall forming a 'roundel within a roundel'. Arched brickwork from the fires to the kiln wall, or timber rafters, create a kind of inverted cone-shaped plenum chamber under the drying floor.

Most of these 'hopper' type kilns have now disappeared but a sectioned brick example has been preserved in the oast which forms part of the Wye Rural Museum at Brook, Nr Ashford.

HOPPER KILN
'Roundel Within a Roundel'

Hopper Type Construction

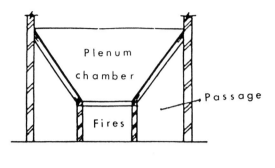

SECTION THROUGH KILN

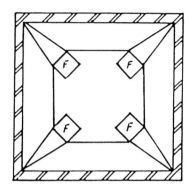

PLAN OF FIRES; TYPE 'B'

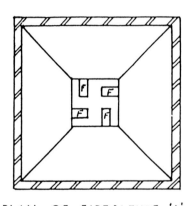

PLAN OF FIRES; TYPE 'A'

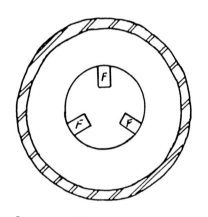

PLAN OF FIRES; TYPE 'C'

Open Fire Hop Curing

Straw or kindling wood is used to ignite charcoal to start the fire. Wood in its natural state is not suitable since the fumes given off taint the hops.

A special Welsh coal 'Anthracite' which is 'smokeless' and does not give off sparks or poisonous fumes is the main fuel. Special checks have to be conducted to ensure low levels of arsenic in the coal before use.

Although these open coal fires were still numerous in the early part of the 20th Century, they have now been almost completely phased out. The primary reason for this was not their fire hazard but to remove all possibilities of hop contamination. Also the arsenic-free coal is costly and difficult to obtain.

COKE FURNACES

Hop contamination was eliminated by employing coke furnaces with only secondary heated air passing through the hops and all the combustion gases kept entirely separated and vented via a flue pipe.

Plant design can be categorised into three different types.

The drying air heated by:
    (i)   a stove erected on the ground floor of the kiln;

    (ii)   a stove in an adjacent building;

    (iii)   steam from a boiler in an adjacent building.

Examples:-
    (i)   (a)   Cockle stove - Week's 'Pure-air' plant.
            (b)   Shew's 'Economic' plant.
            (c)   Drake and Fletcher's plant.
            (d)   The 'Elkington' air heater.

    (ii)   The 'Sirocco' hop-drying plant.

    (iii)   W. Arnold & Sons, steam hop-drying plant.

Shew's Furnace

Cockle Stove

The earliest type of stove used in hop-drying was the 'cockle' stove.

It was made of cast iron in two sections and derived its name from the fact that the top section resembled a cockle shell in shape. The bottom section was cylindrical with a fire door at the front. The outer surface of the stove had fins to increase the heating surface area. This stove was placed just inside the kiln wall on a brick-built base, which also served as an ashpit.

From the domed top section, two cast iron pipes sloped upwards across the centre of the kiln joining at a brick-built flue on the opposite wall. The flue vented to a brick chimney outside the kiln, so that no combustion fumes could come into contact with the hops and cause contamination. Cold air entered the kiln on the ground floor, passed over and around the stove and pipes, becoming sufficiently heated to slowly cure the hops as it passed through them.

A later model as used in Week's 'Pure-air' hop-drying plant possessed four cast iron flue pipes and was sited in the centre of the ground floor with the combustion fumes carried away in a manner similar to that used in the Shew's 'Economic' hop-drying plant (see 113).

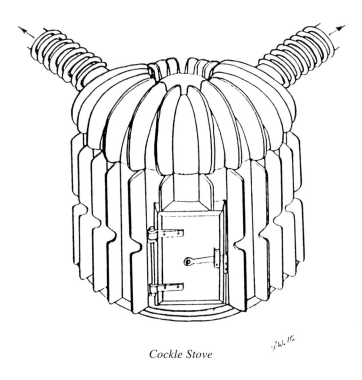

Cockle Stove

Shew's Patent Economic Pure-Air Heater

This plant was manufactured by Messrs Jones & Attwood Ltd, of Stourbridge, Worcestershire.

The furnace stood horizontally on the kiln floor and was made up of cast iron sections; the actual number of sections dependent upon the size of the kiln. The flue pipe was always positioned directly below the centre of the drying floor.

A fire box and ash door were built into the brick wall separating the kiln from the stowage. All firing operations were carried out from the stowage thus preventing any dust or fumes from entering the kiln.

A thin steel casing enclosed the sides of the furnace leaving the top open. The end of the casing was connected to a large brick-built duct, with a wooden or concrete cover, leading to a fan housed in the outer kiln wall. This power-driven fan forced cold air along the duct and around the furnace. A timber trap-door was fitted into the duct cover and operated by a lever in the stowage to admit cold air into the kiln when necessary.

Immediately on top of the flue outlet at the rear of the furnace was fitted a cast iron junction box, from which numerous pipes radiated upwards and outwards towards the outer wall of the kiln, then turned inwards to connect with a second junction box just below the drying floor. A short vertical pipe between the two junction boxes was for support only and did not carry any products of combustion.

The ducted air was first heated by the furnace and then additionally heated by these radiator pipes which were supported either from the floor below or by iron rods or chains hanging from the drying floor above.

From the upper junction box a 9 inch diameter metal flue pipe rose vertically through the centre of the drying floor and terminated in a 'T' junction enabling exhaust gases to vent through the kiln roof just below the cowl.

To prevent the hops from coming into direct contact with the hot flue, a 15 inch diameter, 3 feet high metal sheathing was attached to the drying floor. Any hot air that passed up between these pipes was deflected back onto the hops by a metal flange plate above the sheathing.

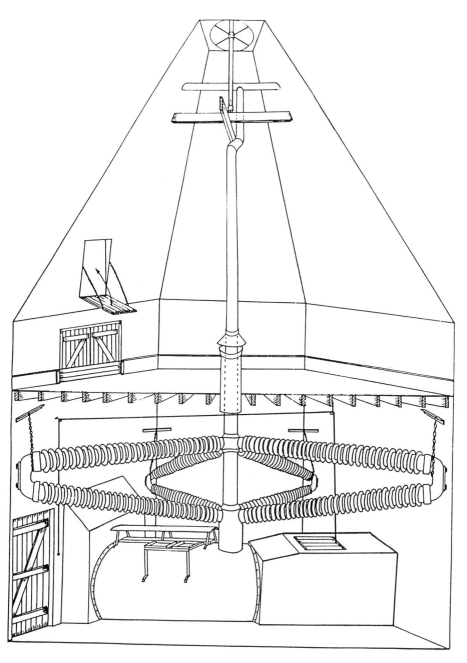

Shew's Pure-Air Heater

R Walter
1983

113

Drake And Fletcher's Hop-Drying Plant

The stove is a rectangular horizontal cast iron type, built up in sections, the number of sections depending upon the size of the kiln. Fins are cast on each section to increase the heating surface. The stove is surrounded by a vertical casing of sheet steel, supported by a framework of angle iron and standing clear of the sides and back of the stove. A space is left between the lower edge of the casing and floor for the admission of air. This casing is fitted flush with the front of the stove and is cut away to give access to the fire and ash doors. Above the front and back of the stove the casing is fitted with doors through which additional cold air can be admitted to the drying floor when necessary. At a height of about 10 feet from the kiln floor the casing is bent sharply outwards, on all sides, to meet the wall of the kiln. The result is that no air can reach the drying floor from below except that passing through the casing. The air, after being warmed by contact with the stove, is heated still further by two flue pipes. These pipes zig-zag upwards inside the casing to give maximum heat transfer and then separate and encircle the inside of the kiln, supported by steel hangers from the floor joists above. Finally they combine into a single flue pipe, which vents through the kiln wall and culminates in a steel cowl or 'T' piece.

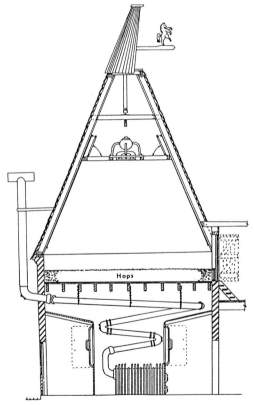

Drake And Fletcher's Hop Drying Plant

Mechanical Stoker

This is a device designed to deliver a continuous unattended supply of fuel (coal/coke) and air to the furnace.

The fuel is fed by means of a worm (coarse pitched) screw from the storage hopper along a feed tube and forced up into the grate. Air is blown into the air chamber of the furnace by an electric fan.

At Saynden Oast the three kilns each had one of these units installed, which were manufactured by Prior Stokers Ltd, London/Glasgow, coupled to a Drake and Fletcher's furnace.

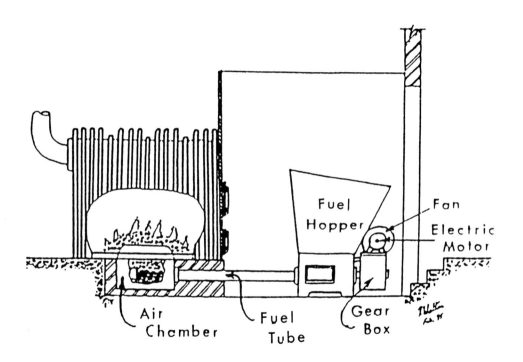

Mechanical Stoker

The "Elkington" Air Heater For Hop Drying

This type of hop-drying plant was manufactured by The Dover Engineering Works Ltd, Dour Iron Foundry, Dover, Kent. A rectangular cast iron stove was mounted on an ashpit located centrally on the ground floor of the oast kiln. Cast iron flue pipes were connected to the top of the stove at each end. These pipes extended outwards and upwards below the drying floor and jointed externally before venting through a vertical flue pipe.

Above the stove a vertical rectangular sheet metal duct passed through the drying floor protruding to a height of 4 ft. This upper section was encased in a wooden 'box like' structure to protect the drying hops from direct contact with the hot metal surface. The top of the duct could be closed during the drying process by a cover that was cord operated from outside the kiln.

Attached to the underside of the duct are two parallel bars, holding a sliding plate (shutter). A moveable woven asbestos curtain hung from the plate, which could either partially or completely cover the top and sides of the stove. This mechanism controlled the amount of heat being directed either through the hops, when the curtain was drawn back (duct closed) or vented into the roof space above the hops when the curtain was pulled forward enclosing the stove (shutter open).

As soon as the hops were dry, the duct cover was lifted and the curtain drawn around the stove. The hot air passed up through the duct and escaped rapidly via the roof cowl. Shuttered vents at ground level were fully opened to provide cold air to cool the dried hops, before removing them to the stowage floor.

When the next load of fresh green hops are spread on the kiln floor this process is reversed. This cycle is repeated twice a day during the hop drying period each year.

During our 19 years of research we have only found one farm where this technique was used for a short period at the beginning of the 1930's and that proved rather ineffective.

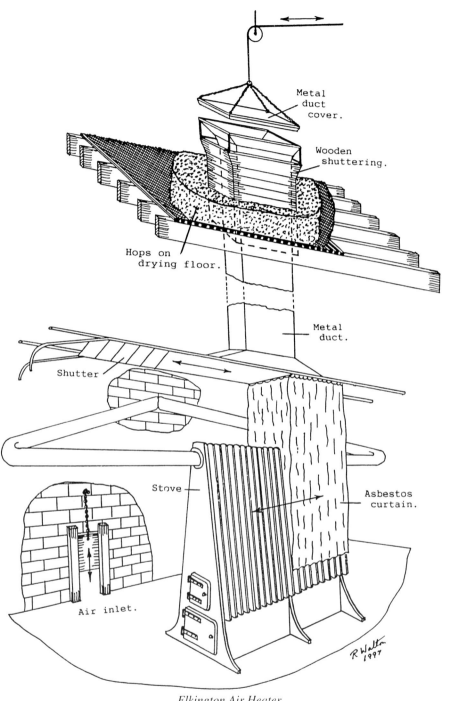

Metal
duct
cover.

Wooden
shuttering.

Hops on
drying floor.

Metal
duct.

Shutter

Stove

Asbestos
curtain.

Air inlet.

Elkington Air Heater

117

THE 'SIROCCO' HOP-DRYING PLANT

Designed and manufactured by Messrs Davidson & Co. Ltd, Belfast, Ireland.

The plant consisted of a multi-tubular heat storage unit enclosed in a sheet steel casing. The firebox was situated in the centre of the heater with a door on either side giving access for cleaning the heating tubes.

Coal or wood was used as fuel with the smoke from the fire passing from the firebox into twenty-five horizontal tubes on both sides and then venting up the chimney at the rear. A fan was placed next to the heater to draw clean air through the heater casing around the heating tubes and then forcing it through the ducts which lead to the spaces below the drying floor. The fan was driven by a pulley belt utilising either a steam engine or tractor drive.

The drying plant was normally housed in a corrugated iron roofed outbuilding.

A 'Sirocco' hop-drying plant was used from the 1920's until 1945 to dry the hops at Messrs Bensted & Gillett's farm, Woodstock, Nr Sittingbourne.

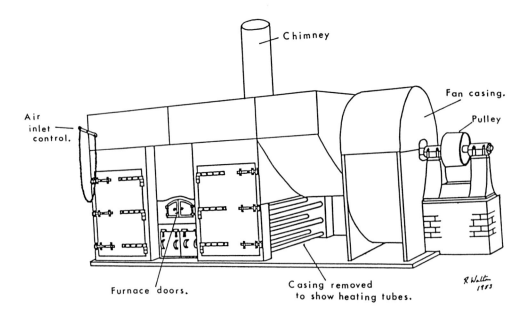

Chimney

Fan casing.

Air inlet control.

Pulley

Furnace doors.

Casing removed to show heating tubes.

118

STEAM HOP-DRYING PLANT

This type of plant was manufactured by Messrs W. Arnold & Sons Ltd, of Branbridge, East Peckham.

A boiler with a battery of steam pipes, a fan and a steam engine were all housed in a brick building adjoining the oast. Cold air was forced through the metal casing, enclosing the steam pipes, by means of the large fan itself powered by the steam engine. This heated air then passed through ducts into the space below the drying floor of the adjacent kiln. As a result a much lower drying floor could be constructed with this method of curing.

OIL FIRING

The early type of oil burning plant consisted of a rectangular tunnel built of fire bricks, enclosed within a metal framework and sited in the old coal hearth opening in the kiln.

An observation hole and ventilation flap were built into one end together with the oil burner nozzle, which forced the hot gases along the tunnel. The fire bricks absorbed this heat and then radiated it throughout the kiln.

This type of construction was later referred to as the 'Coffin' plant, due to its shape and it was necessary to keep a constant watch on the burner's flame.

The present day oil burning plant is housed in the stowage area and not in the kiln and consists of a large horizontally-mounted cylindrical metal casing, lined with fire bricks. A fan at the rear end pushes the hot gases through ducting into the kiln.

In these modern designs there is no necessity for the drier to keep a constant watch on the burner as a 'magic eye' heat sensor automatically cuts off the oil supply should the flame go out.

Although in all these oil burners the actual hot combustion gases pass up through the hops, there is no contamination because the fuel is free of impurities.

'Coffin' Type Oil Fire

GAS HOP-DRYING PLANT

These plants are essentially the same as the early oil burning systems i.e. housed in the kiln but have burners designed to operate on natural gas as the fuel.

Only one hop farmer in Kent is known to currently operate this drying system.

SULPHUR BURNING

At the commencement of the curing process sticks of sulphur were burnt either in an open iron pan or in a sulphur stove generating sulphur dioxide gas which passed through the hops, partially bleaching them.

(1) Open Iron Pan

The traditional method was to ignite the sulphur in the pan by adding two or three pieces of burning coal and then placing it on the ground floor of the kiln.

In recent years the pan was positioned in a fixed metal frame and the sulphur was ignited by liquid propane gas burners (supplied) fed by a cylinder in an adjoining room.

(2) Sulphur Stove

This consisted of an iron fire box supported on iron legs with a small chimney at one end and an access door containing ventilation holes at the other end. Sulphur and burning coal were placed on an iron tray and slid into the fire box. The sulphur dioxide gas passed out through the chimney and upwards through the drying floor.

Note. This practice of sulphur burning was discontinued during the 1980 season, at the request of the Brewers Society.

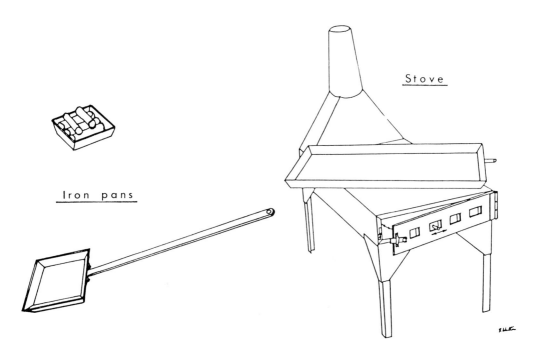

Stove

Iron pans

DRYING HOPS

Moisture Content

The drying of the hops is a skilled technique that requires a careful control of the moisture content throughout the whole process.

The three factors which have to be taken into account are the temperature and flow of the drying air, and total drying time.

Freshly picked hops contain between 75 to 85% moisture and to ensure that they are preserved adequately during storage, they are dried down to a moisture content of about 6%. After drying this level of moisture rises slightly, levelling out at about 10%.

The old experienced drier estimated the moisture content by 'rubbing down' some of the hops in the palm of his hand. If the leaves chaffed finely, leaving no fibrous residue, just a yellow resinous deposit on the fingers, then the hops were judged to be dry enough. The stalks retain the moisture longest and when correctly dried should be brittle.

An instrument called a 'Megger' has been produced to indicate moisture levels. This device has two metal probes that pass an electric current through the hops. The conductivity varies according to the moisture content, which is read from a calibrated scale.

A portable single probe battery operated instrument, designed to measure the moisture content of hops in a pocket or bale, has recently come on the market.

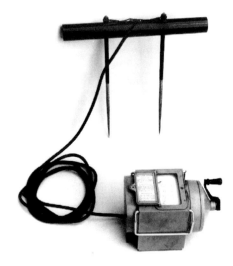

Megger

Drying Temperature

The traditional method of determining whether the hops were drying at the correct temperature, was for the hop drier to place the back of his hand on the hop bed to judge if any alterations were necessary to the furnace.

Temperature monitoring using thermometers was available as early as 1714 when introduced by Dr Fahrenheit, but the skilled driers preferred to trust their personal judgement.

Those farmers who did invest in heat sensors, monitored the drying air temperature at a position about six inches below the centre of the drying floor rather than at the hop bed itself. A freely suspended metal bulb transferred the heat along a flexible metal tube which was connected to a remote thermometer situated near the furnace controls. This could be either a conventional calibrated glass tube or metal dial.

In modern oasts recording thermometers provide a permanent record of the temperature during the drying period.

VENTILATING FAN

Originally, when the only method of hop-curing was by means of ordinary coal fires, the heated air was allowed to rise naturally through the hops. The circular kilns were built because it was thought that they would assist in the efficiency of this natural draught.

The object of installing a fan was to increase the through-put of the oast, by forcing a larger volume of air to pass easily and quickly through a greater depth of hops.

There are two different working arrangements for the fans. The hot air is either 'driven' up through the hops from below the drying floor or 'sucked' up from above the floor.

The base fan is installed in a small brick ducting built on to the outer wall of the kiln at ground level.

A vertical 'puller' fan is mounted in a specially constructed wooden housing which projects from the kiln roof and resembles a dormer in appearance.

Both of these fans were normally driven by a wide flat belt powered from a steam engine or tractor. This would be coupled to a single fan or via an iron counter shaft connecting a number of fans.

A more recent design of 'puller' fan is horizontally fixed to a staging in the roof space, adjoining the cowl support beam. Due to its isolated position this fan had to be driven by electricity.

Nowadays most of the belt-driven fans have also been converted to electrically driven units.

All three of these six bladed fans are similar in size, being approximately 30 inches in diameter.

'Pusher' fan at base of kiln

Steam engine driving 'Puller' fans

19th TO MID 20th CENTURY KILNS

Drying Floor
These floors are 12 feet or more above the fires and about 3 feet below the junction of the kiln wall and roof and are entered through wooden doors from the upper floor of the stowage.

In two-storey buildings (oasts) there is a height difference between stowage and kiln floor levels and therefore it is necessary to climb a short flight of steps to enter the kiln, whereas, in three-storey stowages the two floors are found at the same level.

The drying floor is designed to permit the hot air to pass easily through and consists of large section timber joists spanning the kiln. Quite often, with the larger kilns, a steel or timber beam is placed under and at right-angles to the joists to give extra support.

Wooden slats of 1¼ inches square section with a 1 inch gap between are nailed on top of the joists at right-angles.

Types
(i) A horse-hair mat is spread out over these slats and secured at the edges. This material is used because of its high resistance to the humid, sulphur laden atmosphere. Lifting cloths are hung from hooks on the wall and spread out over the horse-hair mat to carry out the dried hops. These cloths normally number six and were originally made from goats hair.

(ii) In a few oasts with square kilns a roller horse-hair mat is used. The hops to be dried are laid directly on to the mat which is then wound on to a 4 inch diameter wooden roller running the full width of the kiln by means of a cranked handle and cog wheels. The dried hops fall on to a lifting cloth laid out on the cooling floor of the stowage.

James Clifford & Son Ltd.,
Established 1747
Westree Works, Maidstone.

Roller Hair Floor (Square Kilns Only)

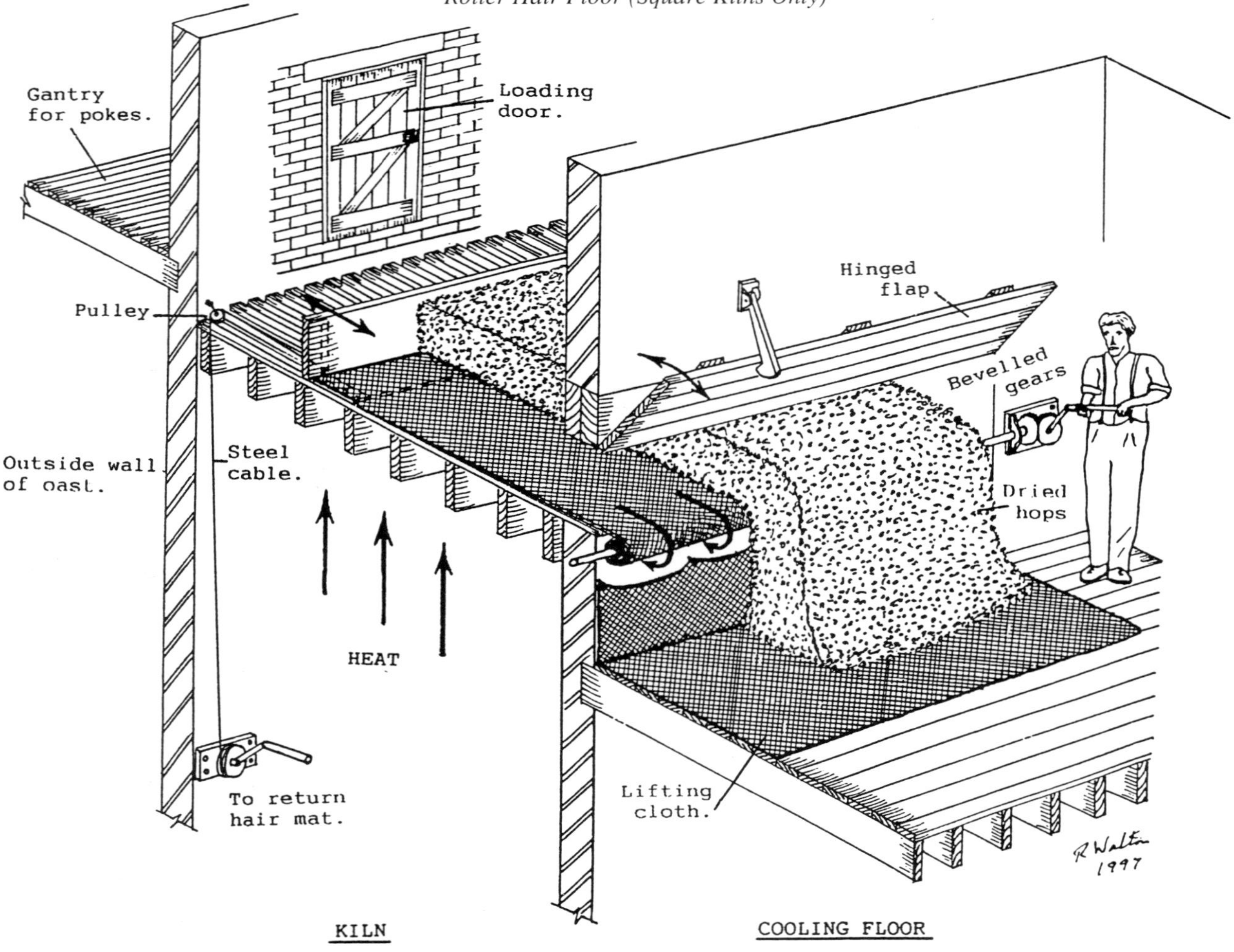

Loading door.

Gantry for pokes.

Pulley

Steel cable.

Outside wall of oast.

HEAT

To return hair mat.

KILN

Hinged flap

Bevelled gears

Dried hops

Lifting cloth.

COOLING FLOOR

R. Walton 1997

126

A small number of square kilns have a movable carriage spanning the floor width on which the hop drier can pull himself, by means of a rope, across the kiln to inspect the drying progress without risk of hop damage.

This carriage consists of one or two planks of timber with a pair of wheels at each end running along rails fixed to the side walls just above the bed of hops.

The roller horse-hair mat and movable carriage techniques were used by Richard Pierce, at 'Goblands Farm', Hadlow.

In the accompanying photograph Jim Houghton, farm foreman, can be seen checking the drying hops.

Carriage, for inspection of hops

TWO-TIER DRYING SYSTEM

In the early part of this century farmers looking for a more efficient way of using the heated air to its full potential introduced the two-tier system.

The timber drying floor was removed from the kiln and a perforated metal louvred floor installed, the perforations allowing the hot air to permeate through the hop bed above. The floor consists of three separate sections per kiln, each operated by individual levers, which open the pivoted louvres and allow the partially dried hops to discharge into removable bins below.

The bins are metal tubular framed structures with cloth linings at sides and bottom, and have an open top. The final period of hot air drying is carried out in these bins. A hinged wooden shutter built into the end wall of the kiln is raised to transfer the bins to the stowage and they are then pulled through on to the cooling floor.

The early system had three large rectangular metal bins per kiln either suspended from rails or mounted on wheels which run in metal channels at floor level. Once in the stowage the hops were discharged on to the cooling floor by opening the hinged bottom of the bin. Later types of bins mounted on casters were rolled out on steel channels from the kiln on to the cooling floor, then turned over on their sides to discharge the hops.

A farmer on the Kent-Sussex border has converted his circular kiln with a very ingenious rotating floor on which six bins, consisting of sectors of a circle, collect the hops in the second stage of drying for transfer to stowage.

Hop Bin

128

Louvred Floor

PROGRESSIVE BIN DRYING SYSTEM

Although this system of drying was introduced in 1963, initially very few hop farmers took it up.

One that did in those early days was Forstal Farm, Lamberhurst, and still uses the system today.

This system is ideal for a single storey large modern industrial type building, where the large area of floor space is available for the layout of the bins.

The hops are fed from the picking machine along a conveyor and discharged evenly into a loading bin, which is either 8 feet square or 15 feet by 6 feet and 3 feet deep with an open top and mesh base, mounted on wheels.

The recently filled trolley bin full of green hops is run on to the end of a track to join from 8 to 24 other bins, containing hops in various stages of drying.

Several of these tracks span the drying floor and below each track there is a large underground plenum chamber (pit), into which hot air is blown from oil burners and

129

fans housed in an adjacent section. This drying air percolates up through the bins and is prevented from leaking by the use of rubber seals along the track.

The volume and temperature of the drying air is controlled for each individual bin by overhead sensors, to enable an even and controlled process.

As the end bin is removed from the track and the contents sent for pressing, a fresh bin of undried hops is added at the other end of the line for progressive drying.

HOP DRYING MACHINES

There are currently two types of hop drying machines being used in Kent and only one of each now remains. Both are of German origin namely the WOLF and BINDER machines.

On leaving the hop picking machine the green hops are transferred by conveyor belts to a holding bin situated in the adjacent drying machine building. From here the hops are mechanically fed onto the top belt in the drying machine, which is the first in a series of three horizontal endless stainless steel mesh belts.

The Wolf machine is 45ft long 10ft wide and 12ft high and when positioned with hop bins has an overall length of approximately 100ft. It is loaded with hops to a depth of 18 inches and each belt remains stationary for a period of 2 hours during the drying process.

However the Binder drying machine operates differently, in that the hops are continuously loaded to a depth of 6 inches onto the top belt. They then progress slowly, being conveyed back and forth along the three endless steel-mesh belts within the rectangular steel casing of the drier. This enables the most economic use of the heated drying air.

In both cases the hops are eventually discharged at the base of the machine. The time spent in the drier can be regulated by controls on the console.

After leaving the machine the dried hops are carried by an elevator into a conditioning bin for a period of 1 hour, before being conveyed to the hydraulically operated baler for pressing into 200lbs bales (87kg).

An outbuilding houses the oil fired burners from where the hot air is fed into a 'Heat Exchange Unit' (HEU). Clean air is blown over the heated ducts in the HEU and the resulting 'Pure Air' then enters the base of the hop drier at 150°F (65°C).

The moisture laden air (reek) then escapes from the drier via metal vents, three or five in number, protruding from the roof of the building.

Wolf Hop Drier

Building housing Binder Hop Drier (3 Vents)

HOP POCKETS

The cured hops were packed into large jute sacks, known as pockets, which when full are 2 feet in diameter, 6 feet high and weigh about 180 pounds.

In the 18th century every hops grower was legally obliged to give notice to the exciseman, on or before the 1st September of each year, the number of acres he had under cultivation, the situation and number of oasts, the place of bagging and all store rooms.

No hops would be removed from store until they had been weighed and the pockets marked by a revenue officer. The weight, grower's name and address and the year of production were all recorded.

The Hop (Prevention of Fraud) Act of 1866 required the grower to mark each pocket in letters 3 inches high, with his name, parish, county and year of production.

The majority of stencils are fashioned from zinc because of its resistance to rusting and lasts the full ten years. Some copper stencils are to be found, but are not very common and were more expensive to produce.

Copper Stencil
Date unknown

Marked Pocket
1880–89

Zinc Stencil
1910–19

Today's modern pockets are constructed of polypropylene and are supplied ready stencilled by feeding them under a rotating master drum.

132

FILLING POCKETS

Until the mid-19th century the pockets were suspended through a circular hole cut in the cooling floor of the oast. A quantity of hops were scuppeted into the pocket, then a man climbed down a short ladder into the pocket to compact the hops by treading on them. As soon as each layer was compacted a fresh supply was scuppeted into the pocket until filled with tightly packed hops. This method was unpleasant work for the man treading the hops; the yellow dust given off was very choking.

Treading the Pockets

This sketch, taken from the 'Illustrated London News' published 6th October 1888, shows a young man treading the hops into the pocket.

In the illustration the short ladder, resting against the window, was used to initially climb down into the pocket, and the balance hanging from the beam above the lad's head used to determine the weight of the filled pocket.

An example of the wooden curb, which supports the suspended pocket, can be seen in the exhibition oast at the Bough Beech information centre, see page 158.

In E.J. Lance's book 'The Hop Farmer' - published in 1838 - he refers to a hand lever press and also a Bramah's Hydraulic Press, but neither appear to have been widely used.

It was not until the late 1850's, with the introduction of the mechanical rack-and-pinion press that the pocket fillers obtained relief from this previously rather unpleasant job.

WOODEN-FRAMED PRESSES

The early wooden-framed presses with a cast iron rack-and-pinion mechanism were made by G. Pierson of Hurst Green, on the Kent-Sussex border but the majority were produced by William Weeks and Son, Maidstone, who supplied most of the Weald farmers, whilst H. & F. Tett of Faversham supplied the Canterbury area.

Two designs were used, the less common 'front winder' with the large gear wheel and its handle to the front of the press, and the 'side winder' which was produced in much greater numbers.

Examples of these wooden-framed presses can be seen in Cranbrook and Tenterden Museums, Wye Rural Museum, also the Museum of Kent Life, Cobtree.

'Front Winder'
B Garrett, Maidstone

'Side Winder'
W Weeks & Son, Maidstone

HOP PRESS

A hand press consists of THREE main parts

(i)   Framework - cast iron or large sectioned timber members.

(ii) Pressure mechanism - rack and pinion.

(iii) Pocket support - canvas sling or timber seat.

The press is erected over a 2 feet diameter hole in the wood floor of the stowage area. The open end of the pocket is folded over a wrought iron hoop, the pocket is then lowered through the hole in the floor, the iron hoop fitted into a recess in the floor and the pocket hangs suspended below. Hops are shovelled into the pocket using a scuppet. At regular intervals pressure is applied to compact the hops in the pocket by means of a cast iron or heavy wooden circular block (ram) 22 inches in diameter, which is raised and lowered on a cast iron rack-and-pinion mechanism controlled by a handle normally sited on the right-hand side of the press.

The pocket is supported during the pressing stages by a strong webbing band or chain and webbing sling passing downwards through the floor in a loop and controlled by a handle on the left-hand side of the press.

Alternatively, the side members of a wooden press extend to the floor below and support the pocket on a horizontal wooden member which is lifted or lowered by chains or rope.

When the pocket is full of hops an iron bar is passed through iron loops on the kerb board supporting the pocket. A spring balance is then attached to the underside of the ram and the bar. When the ram is raised the pocket is supported by the spring balance and the weight is recorded, the bar and balance are then removed. The pocket support lifts the pocket to enable the iron hoop to be removed and the opening to be sewn up. The completed pocket is then lowered to the floor below and stowed until required.

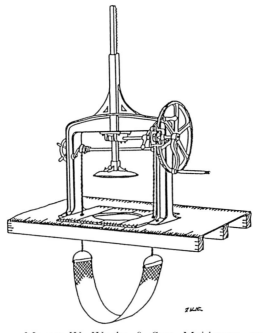

Cast Iron Hop Press

Maker: The Weald of Kent
              Engineering Company,
              Horsmonden

Messrs W. Weeks & Son, Maidstone, produced a similar type of construction, whereas, the Messrs Drake & Fletcher, Maidstone, model had a 'rack' guide bolted to the top of the press and a cranked handle to raise or lower the hop pocket sling.

Weighing the Hops

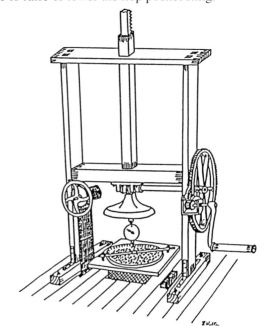

HOP PRESS - G. PIERSON, HURST GREEN

Hop press as used at 'Little Bewl Bridge Farm', Lamberhurst.

This wood-framed press is somewhat unusual in that it is a 'front winder' and also has an iron counter weight attached via a cast-iron pulley wheel, to the ram, which can be quickly thrust down into the pocket prior to engaging the rack-and-pinion handwheel drive to fully compress the hops.

Other interesting features are the timber side members of the press which extend fully to the floor below with a wooden cradle supporting the bottom of the hop pocket during pressing.

The press, along with other artefacts from the author's private collection, is displayed at the 'Museum of Kent Life' at Cobtree, near Maidstone, which opened in August 1984.

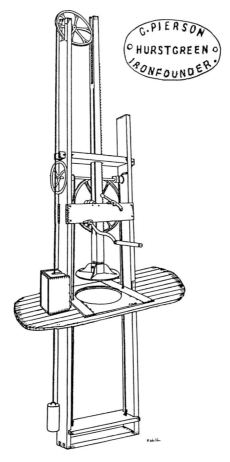

PRESSES - ENTIRELY OF METAL CONSTRUCTION

(i) In the 1890's Messrs. Drake and Muirhead (Maidstone) produced a fabricated rectangular cast iron frame in which the three sections were bolted together. This firm then changed to the well-known Messrs Drake and Fletcher in 1898 and produced a single 'U' shape casting instead.

Drake and Muirhead Press (1940's Power Unit)

(ii) A fine example of a cast iron press (illustrated on page 139) was manufactured by Messrs W. Weeks and Son of Maidstone in the late 19th century and was still in use up to 1980 on Robert Auger's, 'Nashenden Farm', Borstal, near Rochester.

When pressing, the hop pocket is kept in shape by an interesting hinged metal casing positioned by two metal rods and totally encircling the pocket.

The pocket weight is supported by a saddle which is raised or lowered by another rack-and-pinion mechanism.

The securing of the pocket at floor level is unusual in that a metal strap is tightened around the pocket which is folded back over a metal sleeving bolted to the floor.

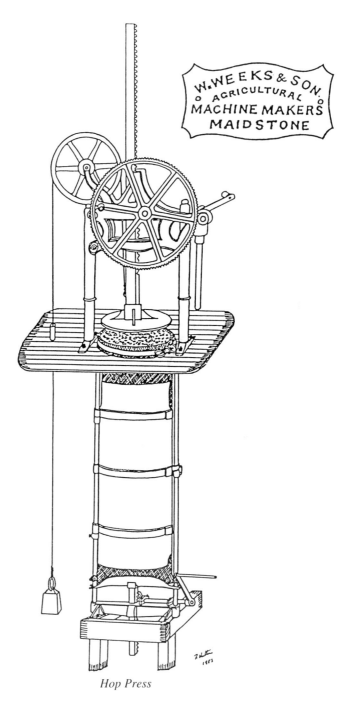

W. WEEKS & SON.
○ AGRICULTURAL ○
MACHINE MAKERS
MAIDSTONE

Hop Press

139

In the 1940's automation was introduced by two Maidstone firms, namely Messrs. Drake & Fletcher, and Kent Engineering and Foundry Ltd. (K.E.F.).

These 'power units' consisted of an electrical motor driving a system of gears coupled to the original 'side winder' drive wheel.

After pressing, the unit drives in reverse to withdraw the ram from the pocket and when clear strikes a cut-out arm to immobilise the power.

Motorised Hop Press

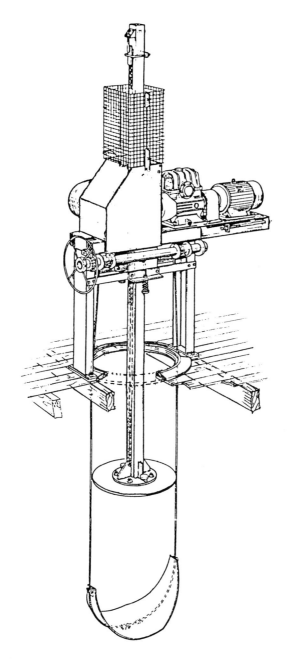

Automatic Hop Press (Heath Engineering Works Ltd.)

141

MODERN BALE PRESS

Hops are scuppetted into a rectangular shaft 1½ feet by 4 feet, the depth varies with the height of the room below.

A rectangular ram covered with a sheet of cloth is forced down the shaft by either hydraulic pressure or a drum and cable mechanism.

The shaft is filled with hops three or four times, depending on the depth, to form a rectangular bale 18 inches deep and weighing approximately 200 pounds.

The sides at the base of the shaft, are either removed or the complete shaft raised to expose the compressed hops. The edges of the base cloth (laid on the base platform before pressing commences) and ram cloth are stitched together using a portable electric sewing machine to form the bale seams.

Pressure on top of the bale is removed and the ram retracted up the shaft leaving the bale to be lifted out of the press.

Each bale is weighed and recorded. A reference number and the name of the farm are stencilled on the bale.

*Hop Baler
(Marsh Engineering,
Ivychurch)*

*Located at
Castlemaine Oast,
Horsmonden.*

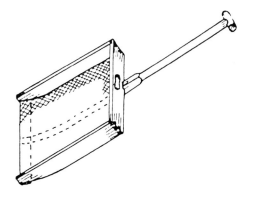

Scuppet

A rectangular wooden frame covered with canvas to which a 'spade type' handle is attached. It is used to 'shovel' the hops into the pockets before pressing.

Hop Pocket Hook

This is a wrought iron handle with two curved hooks, used to manoeuvre the large bulky pockets when stacking or loading.

Hop Pocket Barrow

The full pockets are very bulky and difficult to manhandle, and so are laid horizontally on a special barrow for ease of transportation around the oast.

143

JUST REWARDS FOR THE HOP DRIER'S SKILLS

Since the very beginning of hop production, the drier has always taken great pride in his finished product.

This skill was severely put to the test in the many local competitions arranged by the agricultural associations.

Town Malling Agricultural Association, 1877

A certificate and sum of money (£3 - a princely sum in those days) was presented to the drier, Thomas King, whose farm produced the second best managed sample of hops in the area.

Messrs W.H. & H. Le May, Hop Factors, offered, at the Ashford Cattle Show, a large solid silver Challenge Cup for the Best Quality and Best Managed growth of English hops annually from 1890 to 1938.

The winner each year held the Cup and was also presented with a small personal replica. This Cup was to become the property of the grower winning it three times. So keen was the competition that although many growers won the Cup on a number of occasions no-one succeeded in winning it for three successive years. Hence the Cup still remains in the Le May family to this day.

Challenge Cup

The judging of hop quality is still continued today.

For example, at the Weald of Kent Ploughing Match Association's annual competition, the winner of the various classes offered receives a certificate and small sum of money, with the winner's hops exhibited at the Marden Fruit Show up until 1991. The Fruit Show is now held at the Agricultural Show Ground, Detling.

Hops on Display at the Weald of Kent Ploughing Match

145

SAMPLING

The hop pocket is laid down on its side and about an 18 inch cut made along the seam. The edges of the cut seam are fastened back with iron hooks and a double bladed knife is thrust into the compressed hops, cutting first down and then across the hops to produce a sample cube. The block of hops is removed using special tongs and the faces trimmed on a wooden block. After the sample has been faced, it is carefully wrapped in stout paper to await examination by the buyer.

Hop Sampling

STAFF QUARTERS

Once the hop drying process got under-way on a Monday morning, it was carried out on a continuous basis throughout the week, both day and night.

The drying staff therefore needed sleeping quarters which were provided on the ground floor within easy reach of the kiln fires.

In some of the older oasts this consisted of a bunk and a straw-filled sack partitioned from the main floor area only by a hessian screen.

Modern oasts have a specially built room fully equipped with heating, lighting and cooking facilities.

In "Puck of Pook's Hill" Rudyard Kipling makes mention of two children, who after hop-picking go to the oast to eat hot potatoes fresh from roasting in the kiln fire. There is a delightful line illustration in Kipling's book.

147

LIGHTING

From the 16th century, when oasts were introduced, right through to the 19th century, candles were the main source of light. Due to the obvious hazards associated with timber structures the candles were enclosed by either iron guards or lanterns. Oil lamps were less popular until the discovery of paraffin in 1865, which then lead to the famous 'Hurricane Lamp' being introduced in large numbers. Better illumination was obtained from a pressurised system, for example, 'Tilley Lamp'.

Although gas lighting was introduced into domestic dwellings in the early 19th century, it was never used in oasts due to high installation costs and obvious safety implications; paraffin lamps being cheaper and much safer.

With the discovery of electricity as a source of both power and lighting in the late 19th century, many farmers installed an electrical system. On the larger farms electric generators, driven by steam or oil fired engines, were used either to supply power direct or stored in batteries.

The public electricity supply was restricted mainly to the urban areas until 1948, when nationalisation increased the coverage to include rural areas. Even until fairly recently small farms, in isolated areas, have continued to use paraffin lamps for their source of lighting.

'Lister' Engines Used for Lighting and Power at
Three Chimneys Farm Oast, Goudhurst

Fire Assessors Report

With the awareness of possible fire damage, farmers took out insurance on their buildings, and the following is a condensed version of an assessment conducted by a 'John Smith'.

Report on Mr James Ellis's 'Fire Proof Oast' near his house at Barming, Nr. Maidstone, 23rd July 1814.

The oast was formerly a barn and joined onto a stable, both buildings had an external wall of stone and a tiled roof.

Instead of an open coal fire with a brick hearth as was usual for that period, he incorporated a cast iron boiler which dried the hops by a "double operation". First the smoke was conveyed through a brick flue around the sides of the kiln in the manner of a hot house and discharged itself via a brick shaft on the outside of the building.

The second part of the operation involved the steam from the boiler which was conveyed around the hop floor by earthenware pipes raised a few feet off the floor and the "conjunctive heat" arising from these two operations was expected to be sufficient.

Another principal feature in this kiln, distinguishing it from any other, is that neither wooden joists nor hair cloth were used to dry the hops. Instead of these they were laid on square tiles, perforated with small holes like a strainer, and supported entirely by a number of 4 inch brick walls so that nothing of an inflammable nature was present.

Fire Hazard

Due to the building construction and the actual operations carried out in the oast, incidents of damage due to fire are a continual working hazard.

During the past fifteen years there have been several instances of fire damage to working oasts. These have ranged from small roof fires caused by electrical faults in fans, to the complete destruction of the oast.

Three further coal fire reports on oasts belonging to James Ellis have been located. These are sited at Glebe and Farleigh Bridge. Only one is by the same author, John Smith, and the other two compiled in 1822 are by John Whichcord.

Extracts from the fire assessment reports:

East Farleigh Bridge - John Smith, July 23rd 1814.

"It is in length 105 feet and 37 feet broad contained 32 kilns and as many fire places."

The roof is ... "covered with Thatch with a lead gutter to discharge the rain water at each end".

"...a new fire engine has been lately bought and is kept near at hand."

Glebe - John Whichcord, Sept. 28th 1822.

"This oast contains 40 fires, the kilns are small constructed upon the Hopper principle."

Near Farleigh Bridge - John Whichcord, Sept. 28th 1822.

"The oast 'D' is a wooden building covered with Thatch containing 32 kilns."

These documents were kindly loaned by Christine Swift, Antiquarian Books, Egerton.

Effects of Fire

TUNSTALL OAST, SITTINGBOURNE

Originally a terrace of thatched roofed cottages were built on this site in 1791.

In the late 19th century a 18ft diameter kiln was constructed along side the end cottage.

In order to isolate this end cottage, because of the obvious fire risk it would present in the new use, part of the adjacent terrace was demolished to form a 12ft by 15ft stowage.

Chapter 6

Museums and Attractions

SISSINGHURST CASTLE

An unusual combination of kilns can be seen at Sissinghurst Castle which is owned by the National Trust. This oast has many interesting features including currently an exhibition on the cooling floor covering hopping and local history.

During 1756-63 the castle was rented by the Government and used as a Prison Camp to hold French prisoners-of-war. A pictorial view displayed in the National Trust publication on Sissinghurst Castle depicts a perimeter building on the site of the present oast.

The south east corner of the existing oast is actually part of that building, which was probably constructed originally in the l6th century by Sir Richard Baker and consists of red bricks bedded in lime mortar, laid in English bond.

It was during the late 18th century that the six 12 feet square kilns were constructed on the side, comprising of red bricks in compo mortar, laid in Flemish bond.

The two roundels were added during the time of George Neve, the tenant farmer from 1855 to 1903.

Captain Oswald Beale took on the tenancy in 1936 and this has been past down the family line to the present farmer, James Stearns his grandson. In the 1920's there were 40 acres of hops, but due to the fall in demand by the mid 1960's this had halved to 20 acres. The oast was last used to dry hops in 1967.

Originally the heating in the two roundels was by coke burning furnaces. Evidence of their use can still be seen above the bricked up door opening at the rear. Where the flue exit which was just below the drying floor has also been bricked up and rust from the metal flue pipe over the years has stained the brickwork below. Additionally a 'pusher' fan (which was used to increase the warm air flow through the bed of drying hops) can still be seen at the base of the wall. In 1946 oil fired burners replaced the coke furnaces.

The drying floor in this type of kiln is normally constructed at least 12 feet above the open coal fire for safety sake. However with the introduction of enclosed oil burning heaters, the then farmer Stanley Stearns, in 1954 lowered the slatted drying floors to the level of the pressing floor to enable easy loading and unloading of the kiln. This reconstruction work necessitated the removal of the existing short flights of steps and the formation of new doorways.

In 1957 the ground floor doorway into the small south kilns was enlarged and the redundant six open coal fires removed. A hop picking machine constructed by the Kent Engineering Foundry (K.E.F.) Ltd, Maidstone, was installed in their place.

During the early part of this century a Drake and Fletcher, iron hop press was installed, which was converted to an automatic electrically driven machine in 1946 and continued to be used up to 1967 the last year of hop drying. The present wooden press made by W. Weeks was installed in 1967 but only as an example of the traditional 19th century press formerly installed.

Miss Victoria Sackville-West (Lady Nicolson) lived at Sissinghurst Castle from 1930 to 1962 and her book entitled 'The Land', published in 1927, contains several poems about hops, hop gardens and oasts.

Sissinghurst Castle Oast

BELTRING FARM HISTORY

Mr E. A. White (1844-1922)

Edward Albert White was born in the Parish of Yalding, near Paddock Wood, in early 1844, into a local farming family.

Not only did Edward build the grand oasts at Beltring in the 1890's as a permanent reminder to a once thriving industry, but also produced a consistent and much envied high standard of finished dried hops.

This photograph, taken in 1907, shows him proudly displaying his prize hops, at the foot of the steps to Guests oast, accompanied by a pocket showing the various awards gained between the years 1895-1907.

In May 1908 Mr White was one of the leading campaigners in 'The Great Hop Demonstration' which was concerned with the import of foreign hops. The influx of hops from the continent was threatening the livelihood of the Kent growers who were demanding government action by levying a duty of 40 shillings on imports.

After selling the farm in 1920, he moved to London, where he died in May 1922 at the age of 78 years.

Mr A. Davis

Arthur Davis, farmworker at Whitbreads Beltring Hop Farm, was awarded the British Empire Medal in the 1984 Queen's Birthday Honours list. He was born in 1903 and started working for E.A. White in 1918, 2 years before the farm was sold to Whitbreads.

Arthur has a remarkably good memory of over 70 years of history with his involvement at Beltring. During that time he has dried hops in the Guest Oast, Bell 2, Bell 5 and has seen the change from coal fires to coke furnaces then in the 1950's, to oil firing. He has also witnessed originally the use of natural convection air currents to carry away the 'reek', through to the use of belt driven fans worked by steam or diesel engines, and ultimately to the modern electrically-driven fans. In the early years lighting was by Hurricane lamps, but later on a generator was installed to give electric lighting and power. This inevitably led to the equipment being powered off the National Grid. Arthur recalls the hurly-burly of hundreds of hop-pickers and staff harvesting the 200 acres of hops in the early days and in contrast the relative quiet setting, when in 1953 the Bruff hop-picking machines were introduced.

He also remembers the interests that Mr White had outside just simply growing hops. His 'Spimo' factory was sited on the opposite side of the road to 'Bell Field', where chemical experiments were carried out by Jabez Guest, who lived in 'Brookers House'. The work involved perfecting and manufacture of a hop wash and an horticultural spray product called 'Abol', for the abolishment of pests. When Whitbread purchased the hop farm they sold off the Spimo factory to the local firm of Cooper-McDougall, who in turn sold out to Plant Protection Ltd, later becoming a division of I.C.I. and now more recently Zeneca Agrochemicals.

Theatrical Stage

At weekends and at the end of the picking season, artists were brought down from London to entertain the hop pickers. The stage where the acts took place was sited at the end of Bell 4 (centre of photograph) and it had to be removed in 1945 to make way for the permanent poke storage gantry. A large marquee tent was then erected to continue the entertainments.

Brookers Oast, Beltring.

Built in the mid-1800's, the stowage, of brick construction, had an open ground floor at the front for easy access, with iron pillars supporting the timber-framed and boarded upper storey. The window openings originally had vertical wooden bars with hinged shutters.

By the late 1800's the oast had five round kilns; the three original ones had timber roofs covered with hand-made Kent peg tiles and the later two were constructed of brick throughout. These brick roofed kilns were reputed to be the tallest in the county, but in about 1955 it was decided that they were becoming unstable and were demolished. The hops were dried by using small coal fires, served from a central passage running through the kiln. Later a 'Cockle' stove was installed in the front kiln (see chapter 5). A wooden hand-operated press was sited at the east end of the stowage, close to the front wall. The oast was last used for drying hops in 1948, after which the cowls were removed and replaced with glass domes.

In 1982 Whitbread decided to open the farm as a tourist and education centre. Brookers oast was converted into a working craft centre, involving various materials that included wood, glass, metal, wool and cotton in the upper storey. On the ground floor adjoining the kilns a pottery workshop was established with its furnace where once the old cockle stove stood. Birds of prey were housed for a short period of time on the ground floor at the east end.

The cowls were replaced in 1991 when the oast was fully converted into a restaurant and this remains its present day use.

Brookers Oast (1980's)

156

Recent Use Of Bell 1 - 4

Since 1982 the Victorian Oasts have displayed many Agricultural implements and craftwork. Also they have provided visitor refreshments.

Over the last few years the oasts have been used as follows:

Bell 1 - Hadleigh's Conference and Banqueting Suite.
Roundels Restaurant,

Bell 2 - Pottery Workshops.
Demonstrating - Hand building
Throwing
Slip casting

Bell 3 - Authentic reconstruction of 'The Hop Story'
Traditional hop farming, highlighting the lives and times of the people who lived and worked on Beltring Farm.

Bell 4 - Rural Museum - General agricultural tools and machinery including hopping.
Also a Coopers shop.
The Queen Victoria's Rifles display (they were billeted in the oasts during the Second World War).

Bell 5

Set aside from the main group of oasts is a rectangular brick building with five cowls constructed along the ridge. Built in 1936 it is referred to as Bell 5, as it was the last oast to be constructed on Bell Field. Each of its five kilns were equipped with both a Drake and Fletcher coke furnace and a roller hair mat. Manual cast-iron 'Weeks' presses were used initially and were motorised at a later date.

For 22 years (1973-1995) the oast was leased to a firm which manufactured firstly 'Hopfix' and later 'Hoplets' - a lupulin enriched hop-powder distributed in vacuum packed pellet form to the brewers.

The entire Whitbread Hop Farm tourist complex was offered for sale in the Spring of 1997. The Brookers Oast / Farm House Restaurant complex has not been included in the Bill of Sale.

The new site owners BM Investments will continue with the farm's traditional activities and retain the shire horses. They are planning a number of new initiatives – including an adventure playground in the empty Bell 5 Oast, hop festivals and a micro brewery.

BOUGH BEECH OAST, SEVENOAKS

In 1981 the Kent Trust for Nature Conservation established an Information Centre in this oast, which is leased from East Surrey Water plc and sited at the northern end of Bough Beech reservoir.

The oast comprises of a 25 feet by 20 feet stowage built in hand made red bricks laid English bond in lime mortar, under a tiled roof. The north end has a barn hip roof and it would appear from the roof members that at some time in the past a rectangular kiln had been constructed within the building, with a lath and plaster lined roof.

A 16 feet diameter kiln has at a latter stage been built on the South end. It is constructed of different size red bricks laid in Flemish bond and a slate damp proof course has been built into the brick plinth courses. A Welsh Arch, (dovetailed brick), has been placed above each of the air inlet openings. The roof members have been sprocketed which gives the tiled roof a Bell shape appearance at the eaves.

An interesting configuration is displayed on the cowl vane, depicting Hop Tallying, which can be seen in detail on page 47.

The kiln displays hop artefacts and hop picking photographs on two levels.

A Weeks wooden hop press is sited on the stowage floor and a second press, manufactured by B. Garrett, Maidstone, can be seen in the barn adjacent to the oast.

158

MUSEUM OF KENT LIFE, COBTREE

The late Sir Hugh Garrard Tyrwhitt-Drake, who was High Sheriff of Kent in 1956-57, Mayor of Maidstone twelve times between 1915-49 and director of Style and Winch Ltd, the Maidstone brewers, among many such posts, bequeathed the Cobtree Manor Estate to the Maidstone Borough in 1966.

A derelict 19th century oast was on the estate at Sandling Farm, lying between the River Medway and the M20 motorway, adjacent to Allington locks.

During 1984 it was restored with the intention of establishing an educational museum dedicated to Rural Life in Kent.

The venture was originally set up jointly by the Kent County Council, Maidstone Borough Council and the Cobtree Charity Trust.

In April 1993 the Museum became an Independent Charitable Trust. The site of the old Maidstone Zoo, (in fact Sir Tyrwhitt-Drake's private zoo), on the opposite side of the road from the museum, has been utilised as a golf course, parkland garden and picnic area.

Mr George Brundle, born in 1903, still lives in the small farm house on the site having taken over the tenancy of the farm in 1925. Throughout his 56 years of farming no hops were actually grown on the farm. The oast was used for the stowage of bales of straw, fruit boxes and general farming equipment.

The oast originally comprised of two 16ft. diameter roundels and two 18ft square kilns, the latter being built unusually within the 60ft by 40ft stowage. The walls are constructed of Kentish ragstone and the roof covered with hand-made clay peg tiles.

In 1935 the two square kilns were demolished and a fire in 1951 destroyed the original stowage roof, which was temporarily covered with corrugated iron sheets until the renovation of the oast in 1984. At that time one of the square kilns and the stowage (King-post trussed) roof were both reconstructed.

The oast was brought back to life again using all the old traditional methods of drying and pressing when hops were reintroduced in 1985. The 'Friends of the Museum' planted Fuggle hops in earth mounds, with poles to support them. The hop area was increased in 1986 with the erection of two wire support hop gardens.

On one weekend each year in September, visitors are invited to hand pick the hops primarily on the Saturday. Over the past few years this has coincided with a CAMRA (CAMpaign for Real Ale) event at the museum. The green hops are dried on a traditional slatted floor in the oast over a coal fire during the evening and on the Sunday pressed into a pocket (or two!).

Sandling Farm Oast (Cobtree), before & after renovation

WYE RURAL MUSEUM, BROOK, NR. ASHFORD

This oast previously formed part of the Wye College Agricultural Museum at Court Lodge.

The College, part of the University of London, had a Department of Hop Research until 1987.Currently on the campus there is a Hop Research Unit being part of Horticultural Research International. Dr. Peter Darby is the Unit Head and he with his colleagues, after many years of research, registered in 1996 three new varieties of Dwarf Hops, being the Worlds first.

In early 1997 Wye College sold the museum at Court Lodge and it is now administered by the Wye Rural Museum Trust.

The Museum comprises essentially of a late 14th Century manorial eight isle barn and a 19th Century oast. Both buildings display many irreplaceable agricultural implements of past ages

The oast has a single roundel that was built in 1815. This is rare as the majority of oasts before 1830 had square kilns. The fire chamber is unique in that it has four open brick inner circle furnaces. These were stoked from a passage formed of brickwork arched against the inside of the kiln wall. An internal structure of this type forms a 'hopper' i.e. an inverted cone-shaped plenum chamber under the drying floor.

161

BRATTLE FARM MUSEUM, STAPLEHURST

This farm has a single square kiln oast built in the 1870's and last used to dry hops in the 1930's.

Interestingly the external doorway into the kiln was enlarged during the early years of the 2nd. World War, to house an elephant from a local circus and evidence of this enterprise can be seen in the different coloured brickwork (probably from the local brickworks) used to reconstruct the entrance to the original size.

When the present owners, Brian and Anita Tompson purchased the farm in 1952 they used the building initially to house pigs and chicken.

In 1975 they decided to open the oast as a rural museum. A fire in 1983 destroyed part of the building and some of the collection. This disaster did not quell their enthusiasm to collect bygone items.

Today this unique, fascinating museum has something to interest all ages. The collection housed within the oast and several barns, includes vintage tractors and implements, hop growing tools, hop pickers ephemera and oast equipment. Much of the equipment was manufactured in Kent by such firms as Drake and Fletchers, D. Garrett, and W. Weeks (all of Maidstone) and also Tetts of Faversham.

Oxen where frequently used in the Hawkhurst area for ploughing and pulling the poke wagons from the hop fields to the oast. A pair of oxen, "Stuff" and "Nonsense", are at present working on the farm during the summer months.

The museum is currently (1997) open on Sundays and Bank Holidays from Easter to October. Groups are welcome at any time by prior appointment.

Brattle Farm Oast

PARSONAGE FARM, BEKESBOURNE

On this site near Canterbury there was a brick built four storey Victorian oast. It consisted of a 40ft by 20ft stowage and two 20ft square kilns and was last used to commercially dry hops for brewing in 1989. For the next seven years it was used to dry hops and flowers for decoration, but over the last decade it has had a chequered history.

Its traditional white cowls were replaced with the beehive (square ventilator) design in the late 1970's. Unfortunately during the hurricane of 1987 one of these ventilators was damaged and blown off the building.

Originally the kiln drying floor consisted of slats and a horse hair mat, but in more recent years the mat was replaced by a fine wire mesh. A wooden press manufactured by Tetts of Faversham was used to fill the pockets.

During the period of decorative hop and flower drying the oast was maintained in full working order and displayed the traditional hop tools and equipment used in the trade over the past 100 years, including the Tetts press.

A small hop garden was retained and flowers also commercially grown on the farm. The produce from both these gardens was kiln dried using a natural gas burner then stored under controlled humidity to retain their quality, thus enabling an all year round supply of decorative hop bines and flowers.

Unfortunately in August 1996 fire destroyed the oast and all its contents and only the burnt out shell remains.

The family still specialises in growing and drying hops and flowers for decoration and can be found relocated in traditional farm buildings next door. It is intended to replace the Hop Museum in the future, so that the family can continue to provide education/information on hop picking past and present.

EVEGATE MANOR FARM

Evegate is situated at Smeeth near Ashford, in the area of the County that grew hops prolifically in the 16th century.

It was Reynolde Scot, a hop farmer of Smeeth, who in 1574 wrote the first book about hop farming and oast construction in England.

The present oast at Evegate was built in the early 19th century and consists of two ragstone and brick, 16feet diameter roundels with tiled roofs and traditional cowls. The stowage has ragstone and brick ground floor walls, a timber framed upper storey covered with weatherboarding and King Post roof trusses supporting a slate roof.

A restaurant is currently housed on the ground floor of the oast and an Art materials and Exhibition Gallery, including a picture frame maker, in the upper storey.

Within the farm buildings are many craft workshops and businesses.

In 1996 a Conference Suite and Exhibition Centre was opened. A wide variety of craft hand tools are displayed including hop farming items from the author's collection.

"DARLING BUDS OF MAY"

The novelist H.E Bate's book, "The Darling Buds of May" about the Larkin family at 'Home Farm' made famous in the now familiar TV series was filmed by Yorkshire Television at Pluckley, near Ashford.

The original oast roof consisted of locally hand made Kent peg tiles, many of which were either broken or missing.

In one particular episode of the TV programme the oast roof was supposedly stripped and retiled although it had actually been done six months before. The tiles were replaced with approved modern tiles of a very similar appearance, and the owner of the farm kindly donated one thousand of the original Kent peg tiles to the Rotary club of Ashford, who organised a sale of promotional packs, the proceeds of which were used for local charities in the East Kent area.

Chapter 7

The Greater London Area

Hops were grown in the 19th. and early 20th. century in the Bexley and Bromley areas which were then in Kent. With the expansion of the metropolis they have, since 1965, been absorbed into the London catchment and are currently within the Greater London Boroughs.

In 1844 the hop acreage for the following parishes were;

Bexley 3, Crayford 13, St Mary Cray 19 and Orpington 35.

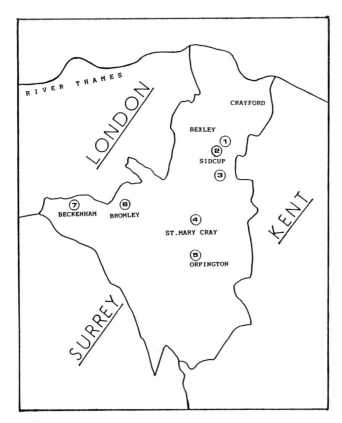

Key:

1. Hurst Farm
2. Halfway Street Farm
3. Ruxley Farm
4. Manor Farm
5. Mayfield Manor Farm
6. Palace Farm
7. Langley Court

THE VINSON FAMILY

The Vinsons were a very active farming family in Kent.

In Kelly's Directory of Kent of 1911 eleven members of this family were listed as individual farmers in various locations which included Belvedere, Bexley, Sidcup, Orpington, St Mary Cray, Swanley, Wrotham and Maidstone.

In Sidcup both Hurst and Halfway Street farms are listed in the census and directories published between 1880 and 1930 as being occupied by a member of the Vinson family. A third farm was located between Halfway Street and Bexley in Day's Lane. Known as Day's Farm, it was shown on the Ordnance Survey Map of 1909 as a single roundel oast opposite the current junction with Annandale Road.

HURST FARM, SIDCUP

The two roundel oast on this farm was sited East of Hurst Place on the A222 Hurst Road, opposite the present junction with Dorchester Avenue.

The oast was demolished in the 1950's having previously been used as Riding Stables during the 1940's and finally for several years as a humble store for the local Horticultural Society.

The cart track which originally lead to the oast has been upgraded and is now called Penfold Lane.

A local resident, Bob Buckingham, who was born in 1917, pointed out the original site where the oast stood and on which old peoples flats have now been constructed.

168

HALFWAY STREET FARM, SIDCUP

The location of the two roundel oast on this farm was on the west side of Station Road, approximately halfway between Holy Trinity Church and the Railway Bridge.

The oast was demolished when the Station Road was widened in the 1930's and a parade of shops were built on the site.

Present day commercial units 135, 137, and 139 are now built on the site of the old oast.

RUXLEY MANOR OAST

At Ruxley Manor, Sidcup, an oast was built in the 19th century onto an existing 13th century Parish Church.

These buildings are currently within the boundary of Ruxley Manor Garden Centre and both the church and the remains of the oast can be observed from the car park.

The oast originally consisted of a single 16ft roundel, with a small stowage connecting it to the church. The latter was also used for the storage of the dried hops.

The oast walls are of brick construction under a tiled roof. A small lean-to building, alongside the roundel, was used for the storage of fuel consisting of both charcoal and anthracite coal.

The church, which is an Ancient Monument, has recently been restored.

Unfortunately the oast is not listed and has deteriorated badly over the last 20 years. The infamous 1987 'hurricane' caused the collapse of the roundel roof.

All that now remains of the oast is the wall of the roundel, which has developed a vertical crack running through the brickwork. Both the stowage and fuel building have been demolished.

MANOR FARM OAST

An oast at 'Manor Farm', St Mary Cray, Orpington, was sited on the main road (A224) south of the railway station, now Oasthouse Way.

Initially it consisted of two brick circular kilns with tiled roofs. A third brick roundel with a rendered brick roof was added at some later date.

Messrs Morphy and Richards started their electrical appliance business at Manor Farm oast in 1936.

The Oast was demolished in the 1930's when Cray Avenue was constructed.

MAYFIELD MANOR

In Orpington high street, near to the corner of Knoll Rise, an impressive brick built oast was constructed at Mayfield Manor having three circular kilns (roundels) at the rear.

This oast was built in 1840 by Mr George Laslett for Joseph Jackson Esq. the owner of Mayfield at the time.

In 1888 the oast, together with two hop gardens, hopper's huts and a cooking house were offered up for sale. At that time Messrs W. & E. Vinson were in occupation of the 122 acre farm.

The oast was partly demolished in the 1920's (roof and upper stories removed) and sometime later converted into offices for 'The Orpington Times'.

Today the site has been redeveloped.

BROMLEY PALACE OAST

The Ecclesiastical Comissioners sold the Bromley Palace estate to Mr Coles Child in 1845.

An oast was sited midway between Widmore Lane and the site of the Bishop of Rochester's Palace, in the area now referred to as Murray Avenue.

Hop fields were sited halfway between the town and Bickley to the east and it is reputed that the farm was the first to supply the London Market with hops each year. This is not surprising since the oast was only 9 miles from the Hop Market at The Borough, East London.

The oast initially consisted of two roundels with steep slate covered roofs and unusually constructed cowls. The cowl's slope was that of the kiln roof with its spindle terminated by a shaped finial, Additionally 'side boards' constructed at the vent opening prevented the ingress of rain.

A third roundel was added in the 1870-80's. Shaped finials were constructed at the apex of all the oast roofs.

The oast was finally demolished in 1927 to make way for housing.

LANGLEY COURT, BECKENHAM

In 1993 the Wellcome Foundation embarked on a programme of landscaping and constructing buildings in keeping with a rural setting at their site in Beckenham.

Two brick buildings in the shape of Oast kiln roundels, complete with traditional cowls, were constructed near the boundary of the site along the South Eden Park Road.

In 1995 the Foundation merged with Glaxo plc. to become Glaxo Wellcome and the eventual fate of these buildings is yet to be finally decided.

Copyright Glaxo Wellcome plc

174

Chapter 8

County Brewers

County Brewers Oasts
There were five major brewers within Kent; three of whom had their oasts located in the mid-west of the county with the other two in the north.

Courage (Eastern) Ltd
The company had three oasts, two in the Yalding area and one at East Farleigh.

A fire during the 1982 drying season destroyed the original modern oast at 'Gallants Farm', East Farleigh. This was a rectangular portal steel framed building clad with asbestos sheeting and having a louvred roof. It contained four square kilns which were oil fired and used a two tier drying system with two electrically driven mechanical presses to fill the hop pockets.

A new single storey steel portal framed oast was constructed on the same site and completed for the 1983 season. The drying process installed was a continuous bin technique, involving the maximum use of mechanical handling, from the adjacent hop picking machine shed to the finished hop bale.

One of the Yalding oasts, at 'Kenward Farm', was of typical Victorian design. This six kiln oast was demolished and in 1984 a new modern oast constructed on the same site similar to that built at East Farleigh in 1983.

The second site, at Congelow Farm, still has the remains of a Victorian ragstone roundel from the original oast. The replacement is a two storey asbestos clad concrete portal framed building, using a two tier drying system similar to that employed in the old East Farleigh oast.

Courage sold these three hop farms to the giant John I Haas, company in 1989.

Two years later the farms came onto the market for sale. In that same year a 'wind blow' just before hop picking time demolished two large areas of hops on 'Gallants Farm'. A local consortium from East Farleigh endeavoured to retain Gallants Farm, in an area that had grown hops for 400 years, but failed to raise the asking price.

In 1991 both Gallants and Congolow farms were eventually sold, but Kenward hop farm was retained. During early 1994, this farm was sold and an arrangement established with a local hop farmer to continue drying his hops in the oast.

Gallants Farm

Trumans Ltd, Laddingford

Upper Fowl Hall oast was originally used by this brewery for hop drying. It consisted of a four kiln oast with brick roundels and a timber-framed weatherboarded stowage. The actual drying process was carried out by coal fires and a metal press used to bag the dried hops. This building has now been converted into a domestic dwelling and renamed 'Willow Oast'.

In about 1948, on an adjoining site, a new rectangular oast was built with brick walls at the back and on the two sides. The frontage had timber louvres running the full width to ventilate the loading area and the roof consisted of steel trusses covered with corrugated asbestos sheeting, with three traditional cowls along the ridge.

A two tier drying system was operated in one of the three 20 feet square kilns utilising oil fires. The modern techniques of pressing the dried hops into bales rather than pockets was used.

The Trumans oast was sold in 1982 after a fire had damaged the roof.

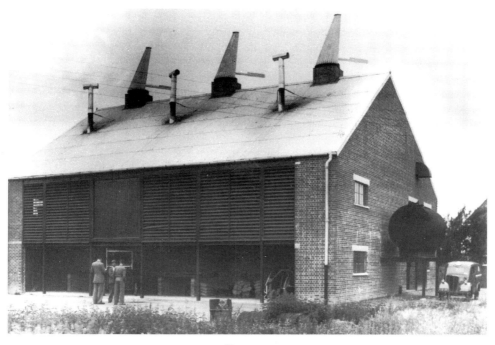

Trumans

Whitbread Hop Farm Beltring

This, of course, is the large well-known site near Paddock Wood (see Page 81). It consists essentially of four Victorian oasts built in 1890'S, constructed of brick and timber under Kent peg tiled roofs. Each oast having five brick roundels, which were originally coal fired and later converted to coke furnaces. The drying system remained the conventional timber slatted floor covered with a horse hair mat and each oast had two Drake and Fletcher cast iron hop presses, which were motorised in the mid 1940's.

In 1942 a neighbouring farm, 'Stilstead', was purchased, with a single four kiln oast on the site, similar to those at Beltring. The coke fired kilns, until the disasterous fire in September 1983, were used to dry the surplus hops when the Beltring oasts were full to capacity

In 1950's oil fired burners were installed in two of the oasts and the mechanical presses replaced with automatic electrically driven equipment. At that time the other two oasts were no longer being used for drying hops.

All four oasts were declared redundant in 1984 and the whole of the Victorian oast complex on Bell Field was converted to a tourist and education centre.

In 1985 a new oast, containing the most advanced mechanical handling and drying techniques, was built at the rear of 'Lily Hoo' house, about a quarter of a mile away. It consists of a single storey steel portal framed building, measuring 240ft by 60ft, clad with asbestos cement sheeting under a pitched roof with six plastic domed ventilators. An oil fired progressive bin system was also installed for drying the hops and a bale press for packaging.

This oast together with about 100 acres of hop gardens were offered for sale by Whitbread Brewery in 1991. They were purchased by a Director of English Hops Ltd, Paddock Wood, who continues to successfully operate the hop farm.

Working Victorian Oasts, Bell Field

178

Guinness Hop Farms Ltd, Teynham

Teynham Hop Farm was originally purchased from a local farmer in 1949. All the drying was carried out in a rectangular four kiln oast which had been purchased along with about 100 acres of land.

During the 1950's additional land was bought and it became necessary to increase the drying capacity of the kilns. This was carried out by converting all four kilns to the two tier system (see Page 128).

In the 1960's further purchases of land were made and the original land, which became isolated, sold back to the farmer from whom it was purchased. The very large acreage left was divided into two units known as Norton and Tickham.

During 1968 a German Binder Drier (see Page 130) was installed at Tickham Farm. Green hops from the three Bruff hop picking machines, working normal daylight hours, were held in bulk bins, enabling the drier to work for 24 hours a day and give maximum drier utilisation. The drying unit was housed in a modern rectangular concrete portal framed building with brick and asbestos cladding. Five cowls protruded from the sloping roof and consisted of small cylindrical metal vents with supported protective conical caps.

Originally an American type timber built press with trunnion and cable drive arrangement was used to press the hops into bales. This was replaced by a metal hydraulically operated bale press in 1970's.

In 1982 the original oast was sold and a larger replacement built on a site at Norton. This architecturally designed, three storey oast consists of a brick built lower structure, with coloured asbestos sheet cladding on the upper storey to be in keeping with the local environment.

The maximum use of mechanical handling has been incorporated to transfer the hops from the picking machines to the drying bins. An uprated 2/3 tier drying system, utilising oil fired burners, has been installed with louvred ventilators built into the roof. A hydraulically operated bale press is used to package the dried hops.

Exceptional facilities have been made available in the oast for all the farm workers, ranging from showers to a fully equipped canteen.

In 1993 both Tickham and Norton farms were offered for sale. The Norton farm was purchased by an established hop farmer and is continuing to operate. But at Tickham farm all the hop picking machinery has been removed, with the equipment being sold at auction.

The German Binder hop drying machine was actually dismantled, with some parts sold to the only other Kent hop farmer with this type of machine and the rest sold as scrap metal!

179

All three of the Bruff hop picking machines were purchases by Hop Engineering, Staplehurst, Kent. One of these machines was shipped to Motueka, Nelson, New Zealand and reassembled ready to pick the 1994 harvest. The remaining two are to follow to different farms in the same area of New Zealand.

Tickham Oast

Norton Oast

Shepherd Neame Ltd, Ospringe

This company purchased its first hop farm, 'Queen Court' in 1944. The farm at that time had a single oast with three square kilns which was sited to the rear of the farm house. Unfortunately, a fire destroyed this oast in the 1950's.

Up until 1984 a second oast in Brogdale Road, Faversham, (about a mile from Queen Court) was used to dry the hops. It was timber framed, clad with weatherboarding under a felt covered pitched roof. The four square brick kilns had the original cowls replaced with louvred ventilators, with additional ventilators constructed in the side of the kiln roof. The hops were dried using oil fires, but two tall brick chimneys remained from the original old coal/coke fires. This oast used the conventional single drying floor method and a wooden motorised press to fill the hop pockets. In 1982 a bale press was installed.

The surrounding gantry on which the pokes were stacked prior to drying, was removed in 1989 when the oast was sold and converted into three domestic dwellings.

Brogdale Oast

Shepherd Neame have built a Public House at Grove Green, Maidstone, in the design of an oast, consisting of one roundel complete with cowl. It was opened in 1989 and is called the 'Early Bird' after the name of an early variety of English hop. One of the hop farming artifacts on display there is a wooden hop press in use until 1985 at Clockhouse Farm, Hunton. The farmer, Mr Peter Day, was one of the last to employ Londoners to pick hops by hand.

Early Bird Public House

There are currently several small breweries in Kent, two of which are;

Larkin's Real Ale
Mr Bob Dockerty, who runs Larkins Brewery at Chiddingstone, comes from a family who have been growing Kentish hops since the 16th century.

The brewery is sited adjacent to the oast on 'Larkins Farm', and was established in 1986.

Goacher's Real Ale
Philip and Deborah Goacher started a family brewery in 1983 at a disused mill, previously producing hand made paper, in the Bockingford Valley, Maidstone.

At one stage they explored the possibility of setting up a brewery and observation gallery opposite the oast at the Museum of Kent Life, Cobtree.

The intention was to use the Fuggle hops grown at the Museum to make a special brew. Unfortunately numerous regulations prevented it from being a viable proposition.

The brewery is now sited at Tovil Green, near Maidstone.

182

Chapter 9

Van Dieman's Land

EMIGRATION OF HOP FARMERS

During the 19th century, many people were emigrating and some of them were Kentish hop growers, who took their expertise with them, including in some cases the hop plants.

One such individual, later to become known as the 'Father of Tasmanian Hops' was a WILLIAM SHOOBRIDGE, born at Tenterden on the 8th July 1781, son of Richard and Ann (nee Standen) themselves hop farmers.

William married Mary Jenkins and they moved to Mottingham, near Eltham in Kent, prior to embarking on the sailing ship "Denmark Hill" on the 21st November 1821, with their eight children.

After lying in port for six weeks, the ship sailed for Van Diemans Land (now known as Tasmania in Australia), arriving on 18th May 1822 at Hobart Town.

Two of their children, William junior aged 7 years and Anne aged 5 years both died of scurvy on the journey. Also his wife and a baby girl born during the long voyage also passed away.

William took with him a letter dated 25th September 1821 from Lord Bathurst of the Colonial Department to the Governor of the Colony. It stated that "he be granted land upon his arrival in proportion to the means he may possess to bring the same into cultivation".

The allocated virgin land of 20 acres was at Providence Valley, Hobart. He had brought with him HOP PLANTS from KENT and successfully harvested his first crop of hops in 1825.

Initially, the Tasmanian oasts consisted of a kiln built on to an existing building, such as, a barn, mill or staging post.

The early purpose built oast kilns were square and constructed of timber wall framing covered with horizontal or vertical boarding and the roof timbers covered with hardwood shingles.

The later very large (30 feet square) kilns had a timber tie member secured across the corners at the eaves level; but the hexagonal and circular kilns had a central metal ring with radiating metal tie rods to give stability to the building.

In 1867 a brick walled hexagonal kiln was built at Bushy Park by William's son, Ebenezer. A plaque built into the kiln wall states "Erected by E. Shoobridge, J.P. in 1867 assisted by his wife, three sons and five daughters. Union is Strength".

An oast with a three storey timber stowage and a 30 feet diameter circular brick walled kiln was built in Valley Field in 1883 by Richard another of William's sons.

In all cases the 'reek' from the drying hops was vented through a louvred timber turret at the apex of the roof. The oasts did not have our typical Kentish Victorian cowls. Hops were dried by 4 to 8 small coal fires, in brick hearths, arranged on the 'hopper' method of construction. (refer to page 108). The hop drying floors consisted of slats covered with a hair mat but no lifter cloths. Large wooden scuppets were used to move the hops from the kiln floor which was at the same level as the stowage floor.

Very heavy hand operated cast iron presses, with a square ram, were used to compact the hops in to bales weighing 250 lbs. These presses were all supplied by Kentish firms Drake & Fletcher, B. Garrett and T. Saveall. They appeared to be a much more robust version of the W. Weeks metal hop press illustrated on page 139.

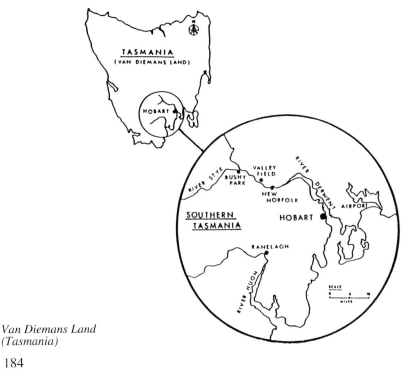

Van Diemans Land
(Tasmania)

VALLEY FIELD

BUSHY PARK

Purpose Built Oasts

185

RANELAGH
Kiln built onto a Barn

NEW NORFOLK
Mill converted into an Oast, now a Museum

186

BUSHY PARK
Modern Oast, 400ft by 200ft
Constructed in 1992

Hop growing still continues, both in the south and north of the Island.

Modern designed oasts have been constructed at Bushy Park and Ranelagh. Mechanical handling of the hops is used throughout, from the picking machine to the bale.

Bushy Park is gas fired whilst Ranelagh uses oil. Both building use the 'Roller Hair' drying floor technique to off load the dried hops.

Chapter 10

Harvest Home

The oast is only used for actually drying hops and bagging for a short, but hectic, period throughout the month of September. For the rest of the year its main use is for storage of fruit and general farm equipment.

The payment to hand pickers is based on the actual volume that each individual attains. The method of recording these individual volumes has altered over the years.

As long ago as the 17th century, in the east of the county a 'tally stick' was used for this purpose. The tallyman would visit the hop garden twice daily and match his numbered master tally stick to the shorter counter tally kept by the picker and then simultaneously cut notches to indicate the number of bushels picked.

The weald and mid-Kent farmers often issued tokens which were made from lead, brass, alloy or thin sheet iron. The earliest examples of these which have been found are dated 1767. Generally a number was either cast or punched on to the face along with the farmer's name or initials. These tokens are found in a variety of shapes and sizes and often a farmer would incorporate his own unique design in order to prevent counterfeit tokens being easily produced.

Since the early part of this century both tally sticks and metal tokens have gradually been replaced by pickers' cards, on which the 'booker' records the number of bushels picked and keeps a master record in a ledger.

The pickers are not actually paid until the end of the season when the various credits are redeemed. The final job of the oast staff is to load the hop pockets on to a horse-drawn cart or lorry and send them off to market.

The working role of the oast is over for another year.

Pokes Heading for the Oast

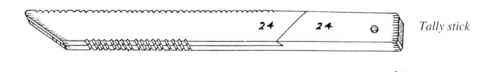

Tally stick

Hop token 1767
Issued by John Toke, Godinton
Great Chart, Nr Ashford

Individual Pickers Card

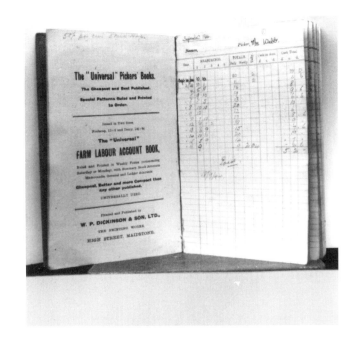

Tally mans logbook

Tallyman and Colleagues - 1910

Paying Hop Pickers
Sketch from 'The Illustrated London News', Published 5th Oct 1867

WOODSTOCK FARM, SITTINGBOURNE

The photograph below taken at the end of the 1921 season shows the successful crop about to leave the oast destined for the London market. The lorry driver is Mr Holness. The gentleman wearing the trilby hat (third from left) is the farm foreman, Mr Streatfield, having just checked off the load. Mr Ranson, the drier, is standing to the right of the foreman. The young 16year old lad standing fourth from the right is an interesting character, His name is Ernest Albert Hopper known to his friends as "Hoppy" and he started work on the farm in 1919 at the age of 14 when it was owned by Captain Gilliat and Frank Bensted who had recently purchased it from Messrs May and Vinson. His father, Ernest, is standing to his left and the foreman's son, Walter, to his right.

At this time the oast had no electricity and hurricane lamps were the source of lighting. Also in the photograph the chimney for the "Sirocco" burner (see page118) can be seen in the upper left hand corner as can the three-louvred ventilators constructed along the ridge of the corrugated iron kiln roof. In the kiln itself a roller hair mat (see page126) discharged the dried hops on to the stowage floor. A wooden press manufactured by S.F. Tetts of Faversham was used to fill the pockets and was sited in the building directly behind the cab of the lorry.

The farm was sold in 1945 to Shell Research Ltd. The "Sirocco" was replaced with up-to-date oil burners and electricity installed on to the site. Hop growing was continued there for experimental purposes up until 1967.

Woodstock Farm, Sittingbourne

Hops Off to Market

TAKE CARE OF THE PENCE

This interesting illustration from 'The Graphic' newspaper represents a gathering which took place on Sunday 22nd September 1878 in the oast at Buffalo Farm, Mereworth, near Wateringbury. The manager of the National Penny Bank spoke about a scheme to open his banks in the hop-gardens at the time the wages are paid.

It was of very limited success, since the London pickers required all their money just to maintain themselves and their families.

Those pickers that did fall ill during the season were often cared for in a local hoppers' hospital. One example of such an establishment is at Five Oak Green, near Paddock Wood, which was an old inn until converted in 1910 for hoppers coming from the East End of London. Now it is a self catering house for the use of churches in three London boroughs.

LAST OF THE VICTORIAN HOP FARMERS

I was strolling along off the beaten track in the heart of the Kentish hop growing region in 1980 and discovered a piece of unique history. It was like stepping back in time.

A Victorian Oast owned by a farmer also born during the reign of Queen Victoria and the way of life had altered very little.

Fuggle hops were grown up poles and hand picked into 30 bushel bins. Twice a day the cry "Hops Ready" rang out through the 3 acre garden. Ben Field (born 1914) was the measurer and pole puller, with the owner acting as tallyperson.

In 1900 the tally was 8 bushels per shilling (5 new pence). In the last year of hop picking at the farm during the 1987 season, the tally was 34 new pence per bushel.

A horse was used to pull the poke wagon to the oast with each poke containing ten bushels of green hops. In 1975 the horse was retired and a tractor took over the job.

The oast had a single 18ft diameter brick roundel, complete with cowl and a 17ft by 20ft stowage. The ground floor walls were also brick, with a timber framed upper storey covered with horizontal weatherboards. Both the kiln and stowage had earth ground floors with a roof covering of tiles.

Three small coal fires in a hopper type construction, (a roundel within a roundel), dried the hops spread out above on a traditional slatted floor, covered with a horse hair mat and goat hair lifter cloths.

Charcoal was used to light the anthracite fires on a Monday morning. A constant vigil was then kept throughout the week by Arthur Farley (born 1907) the drier, who slept in the oast, round the clock only going home on the Sunday. The picking normally lasted 2-3 weeks.

I spent many of my lunch breaks talking to Arthur about his vast wealth of experience in hop drying accumulated over 60 years. I also assisted him in stoking the fires and pressing the hops.

The oast did not contain any complex equipment to measure temperature or humidity. The back of Arthurs' hand was used to ascertain the temperature and the dryness was determined by breaking the stalk.

The hops were pressed into a pocket using "muscle-power", a hand operated wooden press manufactured by "The Kentish Engineering Works Ltd," (previously Drake & Muirhead) later to become Drake & Fletcher Ltd, Maidstone. When full the pockets weighed 180lbs.

The last job was to stencil the season on the pocket, which had already been stencilled with the growers name, Farm and Town. The finished pocket was then stood on a "duck board" so as not to absorb moisture from the ground.

This was to continue until 1988 when the hop garden was ploughed-up and given over to grazing sheep.

During the eight very enjoyable years that I, together with my family and friends, visited the farm at hop picking time, we picked into a bin set aside by the farmer for local charities.

This was the last farm where people were employed to pick the hops by hand. All the present hop farmers in Kent now use hop picking machines, which were introduced in the 1950s.

The last of these remarkable people passed away in January 1994 having devoted their lifetime to traditional hop farming and definitely marked the end of an era.

The Charities' Bin 1986

THE SENTINELS

Stand the oasts in Kents heartlands,

cowl - capped heads against the sky,

time - worn buildings watching, waiting

as the seasons pass them by.

Will Septembers bring their wakening,

horse - drawn wagons to their doors

piled with hops to be unloaded,

spread and dried upon kiln floors ?

Will the Kent and Cockney voices

mingle with the furnace roar;

scent of hops be all - pervading

as it was in days of yore ?

Eileen Dooley
1996

OUR HERITAGE

The main theme of this book has been to record for posterity the construction and use of oasts, from the first recorded oast in 1574, through the various stages of development to the modern day buildings. An additional theme has been to explore the specialised equipment required to prepare good quality hops for the brewer.

Hop growing has been a major part of Kent's agricultural industry for over 400 years. During this period many books have been published about hop growing and picking, but few have made extensive reference to one of the key men, the drier. Our aim was to ensure that his skill was justly recognised.

As with many of the old established local industries, technology and 'progress' have had demonstrative results. Although the number of working oasts are diminishing, the buildings still remain as a unique feature of our county's heritage with conversion to domestic dwellings.

The industry has relied over the years on local skills to improve efficiency and output. Industrial chemists have fought an unyielding battle against insects and fungi in the hop gardens and agricultural engineers have strived to advance the drying and pressing techniques. But recent importation of cheap hops, chiefly from Europe and the shrinking of the export market due to overseas self sufficiency has made it uneconomical for the small local farmer to make a living.

The increased leisure time now available has resulted in more people visiting the country side and strolling along the farm trails, absorbing the beauty and charm of the Victorian oasts, with their distinctive white cowls. This enchanting backcloth has stimulated us over the past nineteen years, and we hope that armed with the information contained in this book, you will also enjoy exploring our county's heritage.

Tallying off

Parsonage Oast, Yalding

199

Appendices

CONVERSION FACTORS

	Imperial Units	Metric Units	
Length	1 inch	25.4	millimetres
	1 foot	0.305	metres
	1 mile	1.609	kilometres
Area	1 acre	0.405	hectares
Weight	1 pound (lb)	0.454	kilograms
	1 hundredweight (cwt)	50.802	kilograms
	1 ton	1.016	tonnes
Capacity	1 gallon	4.546	litres
	1 bushel	36.37	litres

TEMPERATURE COMPARISON

Fahrenheit	Celsius (Centigrade)
140	60
120	49
100	38
90	32
80	27
70	21
65	18
60	16
55	13
50	10

Facts about the 197 farms growing and drying hops during 1984

PICKING	Hand			Machine	
	2			195	
BUILDINGS **OAST**	**Construction**				
	Traditional Stowage, Round or Square Kiln	Rectangular Brick Building. Combined Stowage and Kiln		Rectangular Concrete or Steel-framed Building clad with Asbestos or Metal	
	132	42		23	
	Ventilation				
	Cowl	Louvres		Hinged Shutters	
	121	61		15	
FIRING **(FUEL)**	Coal	Coke	Oil		Gas
	3	1	192		1
DRYING	Hair on Wooden Slats	Roller Hair	2 Tier System	Progressive Bins	Machine
	182	3	7	2	3
PRESSING	Hand operated Wooden Press			Mechanical power driven Press	
	1			196	
	Pockets			Bales	
	190			7	

Facts about the 72 farms growing and drying hops during 1996

PICKING	Hand			Machine	
	0			72	

	Construction				
	Traditional Stowage, Round or Square Kiln	Rectangular Brick Building. Combined Stowage and Kiln		Rectangular Concrete or Steel-framed Building clad with Asbestos or Metal	
BUILDINGS	31	23		18	
OAST	Ventilation				
	Cowl	Louvres		Hinged Shutters	
	27	42		3	

FIRING	Coal	Coke		Oil	Gas
(FUEL)	1	0		69	2

DRYING	Hair on Wooden Slats	Roller Hair	2 Tier System	Progressive Bins	Machine
	44	2	14	10	2

PRESSING	Hand operated Wooden Press		Mechanical power driven Press	
	0		72	
	Pockets		Bales	
	54		18	

OAST DATA SHEET

Ref No.........................Date of Visit….....…...   Authorised by.........................

LOCATION   Map Ref............ Village............   Distance from M/s..........

Address of Farm...

PHOTOGRAPHS.......................................   Maps..

PRESENT USE...   Date...................................

OAST CONSTRUCTION..

STOWAGE Age......   Length...... Width...... Height........ No. of storeys.......

    WALLS: Ground floor........................... Upper floor..............................

    ROOF: Structure............ Pitch...............   Covering….....……...…

    FLOORS: Ground............................. Upper.....................................

    DOORS: Types............................... Ironmongery..........................

    WINDOWS: Types............................... Material…….......…….......

    STAIRS: ....

KILN   Age.......... Number.................   Shape......... Size..................

    WALLS: Material.............................   Bond.................................

    ROOF: Pitch...............   Material.............   Covering….....………...

    DOORS: Type................................   Ironmongery..........................

    VENTILATION: Cowl...........................   Louvre............... Fan…....…..

EQUIPMENT   ....…......

    FIRING   ....…......

    DRYING   ....…......

    PRESSING Type................................. Make...................................

HISTORY   ....…......

    ....…......

Glossary

ANTHRACITE COAL - slow burning yields very little ash, moisture, flame or smoke.

ASHLAR WALL - carefully cut building stone.

BAGGING MACHINE - machine for pressing hops into a pocket.

BAGSTER - a man who trod the dried hops into a bag/pocket before the hop press was introduced.

BAND AND HOOK- long strap hinge consisting of a wrought iron band rotating on a pin fixed to the frame.

BARGE BOARD - inclined timber on the gable of a building.

BARREL BOLT - sliding cylindrical bolt.

BURNER - equipment for the combustion of oil for heating.

CASEMENT WINDOW - window frame with hinged sash.

CENTAL - 100 pounds. Yield per acre.

CHUTE - inclined trough down which hops are passed to a lower level.

COLLAR ROOF - the tie beam is raised above the level of the wall plates.

CONTINUOUS BELT SYSTEM - progressive drying of hops on a moving conveyor belt.

COOLING FLOOR - unheated area in oast where hops are placed to cool before pressing.

COPING - stitching up the open end of a hop pocket.

CORBEL - projecting support on the face of a wall.

COWL - hood mounted on top of kiln to exclude rain and assist extraction of air.

CROWN POST - single vertical timber with four equally spaced wooden struts at the head supporting roof members.

CURING - reducing the moisture in hops so they will not rapidly deteriorate in storage.

DORMER - vertical window in a sloping roof.

DRAUGHT - current of air which passes through the hops during drying.

DRIER - skilled man who is in charge of the oast and responsible for drying the hops.

DRYING - process of reducing the moisture in hops to prevent deterioration during storage.

ENGLISH BOND - the strongest bond for brick walls, consisting of courses of headers and courses of stretchers alternately.

ENGLISH GARDEN WALL BOND - consists of three courses of stretchers to one course of headers.

FAN - revolving wheel with vanes, used in artificial ventilation.

FASCIA - wide vertical board fixed to the feet of the roof rafters.

FINIAL - an ornamental timber projection at the apex of a gable roof.

FLEMISH BOND - alternate headers and stretchers in each course.

FLEMISH GARDEN WALL BOND - a header to every three or four stretchers in each course.

GALLETTED - small pieces of stone in the joints of walls.

GALLOWS BRACKET - framed bracket capable of supporting a load at the outer end.

GANTRY - staging built of timber or steel members, for storage of green hops awaiting drying

GIN WHEEL - large single pulley for hoisting green bags (pokes).

GREEN BAGS (Pokes) - sack of fairly loosely woven material, capable of containing 8-10 bushels of green hops.

GREEN HOPS - freshly picked hops before drying.

HAIR - openly woven cloth of horsehair which is fastened over the drying floor of a kiln covering the wooden slats.

HEADER - brick with the end showing on the face of the wall.

HECTARE - 2.47 acres. Area of hop field.

HERRING-BONE BOND - bricks placed diagonally, used for ornamental brick panels.

HIP - angle formed by the intersection of two inclined roof surfaces.

HOIST - mechanism for raising green bags (pokes).

HOP FACTOR - dual agent of the Hops Marketing Board and individual growers.

HOP FARM - an agricultural unit on which hops are cultivated.

HOPS MARKETING BOARD - authority set up in 1932 to administer the Hops Marketing Scheme, established following the Agricultural Marketing Act.

HOPPER - large wood or steel funnel to hold the hops.

HOP TOKEN - cast or stamped metal token issued by growers, with values from 1 to 120 bushels.

HOVERING-UP - loosening up the hops on the drying floor of the kiln to prevent compaction.

HYDRAULIC - application of liquids to machinery.

KEELE - furnace in an oast

KILN - building used for drying hops by warm air.

KING POST TRUSS - roof truss with a single vertical member from apex to middle of tie beam.

KNAPPED FLINT - snapped flints set in walls, roughly squared on the facing end.

LATHS - strips of wood, small section, for supporting plasterwork, or slates and tiles.

LIFTER CLOTH - loosely woven cloth hung on hooks on the kiln wall, covering sections of the hair. Used to carry the dried hops to the cooling floor.

LOUVRES - inclined wood surfaces that admit light and air but exclude rain, used in Turrets, Doors and Windows.

LUCARNE - vertical window, built within the roof, on the face of the building.

MARKING - record on each pocket name of grower, parish, county and year.

MEGGER - instrument for measuring the moisture content in pressed hops, by their electrical resistance.

MONO-PITCH - single pitched roof.

OAST - building used for the drying, pressing and stowage of hops.

PLENUM CHAMBER - the part of the kiln surrounding the furnace.

POCKET- sack made of closely woven material, into which the hops are pressed after drying

POKE - sack of loosely woven material, containing 8-10 bushels of green hops.

PRESS - machine for pressing dried hops into a pocket or bale.

PURLIN - horizontal beam supported by the principal rafters of a roof truss.

QUEEN POSTS - the two vertical posts in a framed roof truss.

QUOIN - external angle of a wall.

RACK AND PINION - mechanical drive, a toothed rail (rack) engages with a toothed wheel (pinion).

REEK - high humidity produced by rapid water evaporation during the initial stages of drying.

RIDGE - apex of a roof.

ROLLER HAIR- an unloading technique used in square kilns, the hops are unloaded on to the cooling floor as the hair is wound around a roller.

ROUNDEL - kiln, circular on plan.

SAMPLE - rectangular section of hops cut from the side of the pocket.

SASH - window frame in which the glass is fitted.

SCUPPET - wooden 'shovel', the framework is clad with hessian, used for moving the dried hops.

SHINGLE - thin tapered section of timber, normally Cedar or Oak, used for external covering.

SHUTTER FLOOR- kiln floor with horizontal pivotted shutters(louvres) when open allow the hops to fall into the bins below.

SOFFIT - horizontal board fixed to the foot of roof rafters.

SOFTWOOD - wood from coniferous trees, e.g Pine, Fir, Spruce, etc.

SPROCKET PIECE - short wedge-shaped piece of timber nailed to the foot of the rafter to break the surface of the roof

TALLY-STICK-wooden slat used for recording by notches the number of bushels of hops picked.

TOP FAN - power driven fan which extracts air from the kiln above the hops.

TREAD - horizontal surface of a step in a flight of stairs.

TRUSSED ROOF- triangulated frames placed at 10-12 feet intervals along the wall, to carry the purlins.

TURNING PIECE - solid piece of timber to carry the bricks of a segmental arch during construction.

TWO-TIER SYSTEM - drying hops using an upper floor of pivoted shutters.

WALL PLATE- horizontal timber on a wall to distribute the pressure from floor joists or roof rafters.

WATTLE AND DAUB - rough timber framework covered with interlaced wicker work and surfaces plastered.

WEATHER-BOARDING - wooden boards covering the external face of timber-framed buildings.

WELSH ARCH - a brick cut wedge-shaped, like a keystone, supported by adjacent bricks.

WINCH - hoisting machine based on the wheel and axle mechanism.

WINDERS - radiating steps, to form a change of direction in the stairs.

YORKSHIRE LIGHT - solid window frame, one half of which is fitted with a horizontally sliding sash.

ZENTNER - 50 kilograms = 110 pounds. Hop-production/measure.

BIBLIOGRAPHY

Bradley, Richard   The Riches of a Hop Garden, (1729).

Burgess, A.H.   Hops - The Principles of Hop Drying, (1964).

Clinch, George   English Hops, (1919).

Cronk, Anthony   Oasts in Kent and East Sussex, Archaologia
                          Cantiana, Vol.94 (1978) and Vol.95 (1979).

Edwards, J. Morton, B.   A look back at Orpington (1991).
Sign, T and Turner, D.

Filmer, Richard   Hops and Hop Picking, (1982).

Jones, Gwen and   Oasthouses in Sussex and Kent, (1992).
Bell, John.

Lance, Edward Jarman   The Hop Farmer, (1838).

Lawrence, Margaret   The Encircling Hop, (1990).

Loudon, J.C.   An Encyclopaedia of Agriculture, (1831).
                          An Encyclopaedia of Cottage, Farm and
                          Villa Architecture, (1842).

Muggeridge, C.J.   Map of the Hop District of Kent and
                          Sussex, (1844).

Scot, Reynolde   A Perfite Platforme of a Hoppe Garden, (1574).

Scott, Mick   Bromley, Keston and Hayes (1993).

Sutherland, Emma   Hop Pickers Hovels, Huts & Houses (1995).
and Walton, Robin

Index

THE AUTHORS

Robin and Ivan Walton are father and son who have over the past 19 years researched, surveyed and studied the construction and working of Oasts throughout Kent, also compiling of historical records for posterity.

They have photographed and sketched many variations in Oast design and equipment. Additionally they have built up a computer database covering the data collection.

'Kentish Oasts' is the third book they have jointly produced, the others being:

'Oasts in Kent, 16th to 20th century, their Construction and Equipment'(1985),

and

'Beltring Hop Farm, Paddock Wood, Kent, 150 Years of History' (1988).

Robin was educated at the Maidstone County Technical School for Boys and started his working life as an apprentice carpenter and joiner with a local building firm. During which time he gained first hand knowledge of rural building construction. He retired in 1990 from the post of senior lecturer in the Building Department of the Mid-Kent College of Higher and Further Education, Maidstone.

Ivan now lives in the Weald of Kent, being in the heart of the Hop growing area. In the 1980's he went Hop picking on what was eventually to be the last hand picking farm in Kent. He was also educated at the Maidstone Technical School and on leaving school started work at Plant Protection Ltd., Yalding, in the laboratories. He is a Chartered Chemist and currently Manager of Quality Control at Zeneca Agrochemicals, Yalding.

They are both Friends of the Museum of Kent Life at Cobtree, which has a working oast and displays part of their own private collection of hop artefacts.

214